PLAYING WITH PLANETS

Gerard 't Hooft
University of Utrecht

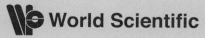

NEW JERSEY • LONDON • SINGAPORE • BEIJING • SHANGHAI • HONG KONG • TAIPEI • CHENNAI

Published by

World Scientific Publishing Co. Pte. Ltd.
5 Toh Tuck Link, Singapore 596224
USA office: 27 Warren Street, Suite 401-402, Hackensack, NJ 07601
UK office: 57 Shelton Street, Covent Garden, London WC2H 9HE

Library of Congress Cataloging-in-Publication Data
Hooft, G. 't.
 [Planetenbiljart. English]
 Playing with planets / by Gerardus 't Hooft.
  p. cm.
 Includes bibliographical references.
 ISBN 978-981-279-307-2 (alk. paper) -- ISBN 978-981-279-020-0 (alk. paper; pbk)
 1. Physics--Miscellanea. 2. Astrophysics--Miscellanea. I. Title.

QC75.H66 2008
530--dc22
                                    2008038545

British Library Cataloguing-in-Publication Data
A catalogue record for this book is available from the British Library.

This English edition is a translation of the Dutch edition *Planetenbiljart*, © Gerard 't Hooft, 2006, published by Promethus/Bert Bakker.

All rights reserved.

Hampshire County Library	
C014655161	
Askews	Dec-2009
530	£14.00
	9789812790200

Printed in Singapore by Mainland Press Pte Ltd.

Preface

As a theoretical physicist, my daily work consists of research and education in highly specialized topics of theoretical physics. Usually, I concentrate on the tiniest constituents of matter, the elementary particles, and the way they interact. It is the experimental physicists who do the marvelous job of designing and operating the complicated constructions with which they make all their observations. They identify the particles and measure their properties. We theoreticians then do our best to put all these findings in a proper context, after which we try to make predictions as to what might still be there to be discovered next, and how this might be done.

The language we use and the formulae we write down, can only be understood by a small number of people. It is a universal language, that of mathematics, and it can be used for the tiniest particles as well as for properties of the universe as a whole, and also the stars and planets that populate it. But it is a very difficult language to learn.

This book is in a different language; that of normal people, who do not necessarily have a comprehensive understanding of mathematics. It is also on a different topic: that of speculations concerning the future. Science fiction mixed with science fact, known by some as "science faction".

My aim was never to lose sight of what we know about science, facts that I believe cannot be ignored when speculating about the future. Still, in spite of all restrictions this imposes, the world that lies ahead of us is a fascinating one. This is the world I wanted to describe, although every now and then, I did let my fantasies run wild.

This book was first written in Dutch, my mother tongue. My daughter Saskia then translated it into English. With great enthusiasm, she produced an English translation that at the same time improved the original Dutch text considerably.

I profited from numerous discussions with friends and colleagues. We specially thank Edward Fredkin for his many interesting comments and suggestions, Joanne Furniss for her editorial contribution, and Annemarie Kleinert for a critical reading of the manuscript.

Playing with Planets (originally *Planetenbiljart*, in Dutch) became a personal testimony of the fantastic speculations and day dreams one can be inspired to, by real science.

Utrecht, August 26, 2008

Contents

Preface v

1 Countdown 1
2 Take Off 10
3 Inside 14
4 Computers 20
5 Paper 33
6 Robots 41
7 Victoriamaris 45
8 A Malleable Earth 49
9 Flying Kites 67
10 The Stars 70
11 The Colonists 81
12 The Cambots 88
13 The Neumannbots 98
14 The Genes 108
15 Pulling Hard 115
16 Aliens 124
17 Playing with Planets 130
18 Idiocracy 136

Websites 141
Illustrations 143

chapter 1

Countdown

MY SPACESHIP WAS painted a brilliant white and sported sleek red and black stripes. The small hatch windows appeared black in color, designed as they were to block the hazardous ultraviolet light emanating from the suns. Take-off and landing were executed vertically. Once on the ground, the spacecraft rested on its small stabilizing fins, to which its three or four engines were attached. You could tell the spaceship had already traveled a great deal; its hull was dented quite a bit. My spaceship had brought me to loads of planets and moons in faraway star systems. The elliptical orbits I chose to navigate around the planets no longer held any secrets for me. I had also learned why it is necessary to protect yourself against the blistering and tempestuous solar winds.

How old was I? Nine or ten perhaps. The spaceship was my own design and it was glued together from paper. The engines were small and needed little fuel because they worked on anti-gravity, a principle I had invented especially for that purpose. Sketches of the planets that I had visited along the way could be admired in my drawing book.

Naturally, an important aspiration during my interstellar expeditions was to stay ahead of all other space travelers. Now this was far from easy. I discovered I had competitors, indeed, many of them. They were the writers of science fiction novels, whose creations were far more imaginative than mine: they invented spacecraft that traveled hundreds of times faster than the speed of light; they met extraterrestrials who nourished themselves with pure thought; and they described aliens who traveled in the comfort of their own home planet, by hopping right into superspace with just a snap of their tentacles. Against such adversaries I would never be able to win.

But I had one consolation, which kept me going. The others were cheating! They modified the laws of nature way beyond credibility. Forcing a wormhole through space and time or paranormal communication — those weren't possible in my wildest dreams. If you don't impose any of the constraints dictated by science, science fiction really isn't that much fun anymore. No, if you want to undertake interplanetary travel, you must comply with the laws of nature — and find the loopholes within. Now *that*'s what's important, because the laws of nature are forbidding; disobedience will never be tolerated. No, you have to be cleverer than that.

Trust me, I know. Because I now know a lot about the laws of nature. I studied physics and have made it my profession. This awesome field is my life's passion. As a physicist, you realize that the laws of nature are not to be messed with. With an amazing mathematical precision, a certain Isaac Newton formulated laws explaining that the gravity of planets, stars and moons is generated solely by their masses, and that this force cannot be influenced by any other external source. One of the conclusions to be drawn is that anti-gravity is impossible, even if you take into account the adjustments one Albert Einstein made to Newton's laws 200 years later. Creating anti-gravity, or whatever else others might conjure up to neutralize the gravitational force, is simply out of the question.

But that's only for starters. There is a lot more that is incompatible with the laws of nature. Indeed, the laws of nature precisely specify what you *cannot* do, even more so than defining what *is* within the realm of possibilities. This has grave consequences. Brace yourself for what comes next:

- It will never be possible to travel faster than the speed of light. Ever.

"Dear Mr. 't Hooft. Surely you have heard of the new Tachyon theory, developed by Gerald Feinberg, and you must be aware of the recently published articles of Luis Gonzalez-Mestres. I am currently developing a brand new design for a faster-than-light machine, and am inviting other like-minded pioneers to invest in my invention. Presently it only exists on paper and hasn't actually been built yet, but ..."

I receive such letters all the time. Gullible investors are guaranteed to lose their money. While velocities greater than the speed of light do exist in physics, spacecraft will never be able to make use of them. Consider a beam of light emanating from a lighthouse. The torch within spins around, and if you stand sufficiently far away from the lighthouse, you can see a spot of light swiveling around with superluminal speed. But it is immaterial; it is impossible to transport any people on that blob of light.

- For every transfer of information a medium is required, such as sound, light or even a piece of paper. Whichever medium is chosen, no message can go faster than the speed of light.

So you cannot even transport a letter with the fast-traveling bundles of light coming from the lighthouse. This feature is shared by all known laws of nature. It is a basic principle that explains a lot about the rules that govern our existence.

- Energy can be transformed into heat, but conversely, only *differences* in temperature can be transformed into usable energy.

Another one of those things: a Perpetuum Mobile, a machine that magically generates kinetic energy out of nothing, is a fabrication. You can't get energy back from heat, but differences in temperature, such as those produced in a steam engine, can be used to generate a lot of energy. This is also the subject of many letters, which end up without much ado at the bottom of the filing cabinet under my desk.

- It is not possible to accurately determine both position and velocity of a tiny particle at the same time. It's either position or velocity!

The mathematical formulation of this law is a bit too complicated for the purpose of this book, but the so-called Heisenberg Uncertainty Principle is of such importance that it must at least be mentioned. The principle poses many limitations to what can be done to atoms and particles.

And so on.

But now what? While science fiction writers might be developing their stories on a load of hogwash, my own paper spaceships didn't work so well either. Is there really no way to travel to the Moon other than in NASA style, with monstrous money-gobbling machines, filled to the brim with fuel — and without my beloved hatch windows?

Well, that might be a bit too hasty a conclusion, as I will explain in Chapter 15: nature's laws allow for another way to lift up into space. But how? That will remain a secret for just a little while longer.

And what about those paranormal phenomena? Aren't the tabloids always full of them? I will take an even more contentious stand: the reason they are called *para*normal is because they are not compatible with the laws of nature. That there are still people out there who attribute any credibility to these phenomena is because they are not taking the laws of nature very seriously. That is odd, considering they owe all their everyday conveniences to these laws, such as their car, television, central heating and what not.

"Mr. 't Hooft, why are you being so cold and harsh? Can't you soften up these laws a bit? Why not allow an exception or two — that wouldn't hurt anyone?."

I also get a lot of those letters. "*Scientists should be more modest,*" I read one day in a letter submitted to a newspaper, "*there are truths other than scientific ones.*" That may be, depending on how you look at it. However, in no way will those alternative truths allow for circumventing or diluting the laws of nature.

Authors of science fiction completely disregard the limitations imposed by those laws. That's why we end up reading about powerful laser beams used for teleportation from spacecraft to the surface of fantastic imaginary planets — marvelous nonsense that lets you escape reality for a while. But while authors certainly create wonderful dreams, even the most outrageous absurdity isn't crazy enough for some writers.

Do you want to enjoy this sort of reading? By all means, go right ahead, read your science fiction and dream on. But remember it is fiction and has little to do with science; not even with the science of the future or with any science known to faraway aliens on distant planets. Most science fiction writers convert the wonderful field of physics into an unrecognizable blur of who knows what, just so that their storyline has at least the appearance of respectability. *"Beam me up, Scotty,"* a man recently sentenced to death exclaimed when he was allowed to say his last words. To no avail; apparently Scotty could not find the button in time.

A negligible minority of science fiction writers attempt to portray a somewhat more realistic version of what the future might hold for mankind. In his 'Mars Trilogy', Kim Stanley Robinson describes how he believes the colonization of Mars will be accomplished. First, robots will be sent to build living quarters for the planet's first colonists. Then, a selection will be made of 50 men and 50 women, representing all continents on planet Earth. Of these, 35 will be American and 35 will be Russian, and each will have a particular specialty or strength. They are 'The First 100', who will undertake the nine-month journey to Mars in an impressively sizable spaceship.

The Mars Colony expands quickly through migration and indigenous population growth. Humanity spreads across the entire planet, which increasingly resembles a suburb of Los Angeles. After only a few generations, the new inhabitants succeed in warming up the atmosphere of the planet and maintain a minimum level of oxygen required to support life, until there is no longer a need to wear space suits. This concept is called 'terraforming', a treasured notion among science fiction visionaries.

Robinson believed that terraforming could begin with building a few windmills, which would generate the necessary rise in temperature. This is rather naïve, to put it mildly. However, his scientific ideas are not nearly as far-fetched as those of other science fiction writers. If terraforming is at all possible, it will take many generations before a change in temperature will be noticeable on a planet such as Mars. Just consider how long it has taken humanity to generate any noticeable changes in our atmosphere on Earth! And it won't be accomplished by building windmills, but with greenhouse gases (I will tell you more about this later).

The speed with which Robinson believes Mars can be colonized — he mentions several waves of immigrants — seems unrealistic to me. The air on the surface of Mars will remain too cold, too thin and too toxic for a long time to come. Robinson paints a pretty picture. However, as I will argue further later on, future inhabitants will have to live either in large glass domes or underground.

And then there are these so-called 'serious researchers', or futurologists, who attempt to base their ideas for the future on scientific findings. But their arguments are equally unconvincing. Because what is their scientific support based on?

Their argument is usually straightforward. *"Let's go back in time, just a few centuries"*, I read recently in a quasi-scientific text *"and ask the scientists from way back then, whether they would have been able to envision our life as it is now, with its cars, airplanes, television, skyscrapers, Internet and numerous medical wonders. They would have been flabbergasted. Is it really too far-fetched to think that the science of the 21st century will astound us in a similar fashion? Like the progression from horse-drawn carriage to airplane, wouldn't the vehicle of the future be similarly advanced compared with our current modes of transport? It would, wouldn't it?"*.

This is as far as this futurologist's argument went. Of course, he could have consulted physicists, engineers and other professionals who understand the laws of nature and technology, and who might have been able to tell him which improvements could be envisioned for the future and what its limits are. But then again, they have been utterly wrong in the past, haven't they: wasn't it Max Planck's physics teacher who said that physics was *"finished"*? Didn't a Lord Kelvin, at the turn of 20th century, admire the *"beauty and clearness of theory [of physics in his days], overshadowed by just two clouds"*? Those two small clouds would develop into major storms: Quantum Mechanics and Relativity Theory, the two pillars of modern physics.

Such isolated and unfortunate remarks continue to haunt current scientists and as a consequence, futurologists continue along the same path as those other storytellers who dominate the world of science fiction. And how on earth can that path be reversed if even celebrities such as Stephen Hawking and Carl Sagan romance producers of science fiction movies with their tales of space warps? Laurence Krauss as well, in his book 'Physics of Star Trek'; how is a level-headed physicist who respects only the true laws of physics going to make it clear to the greater public that much of the so-called 'physics' used in science fiction, or at least the vast majority of it, is only an illusion? Mankind will never be able to travel faster than light — even the speed of light itself is far greater than the velocities that we will ever achieve ourselves. Communication will not go faster than light either, and paranormal communication is out of the question altogether.

You shouldn't compare the current level of modern science to what it was in the late 1800s. During the 20th century, science and technology advanced to such a degree that extrapolation can be made much more accurately now than even a reputable scientist such as Lord Kelvin was able to do more than a century ago; it is perhaps even more unfair to make a comparison with a 19th century physics teacher. *"Why do you say these things,"* I asked one such author, *"surely you realize this goes against everything we know?"* *"Well, yes, I do,"* was his response, *"but if I write that, my books will never sell!"* And so it is. My book will undoubtedly sell fewer copies than his.

But please don't misunderstand; the physics of the future will continue to surprise us and it is possible, even likely, that the future will reveal remarkable technological developments. This potential will be the focus of my book. But we will assume that all the laws of nature presently known to us are accurate, or at least close enough to the truth that we should not expect any major deviations. Contrary to popular belief, the laws of nature known to man a century ago have not proven to be incorrect since. There have been subtle corrections, such as to Newton's laws. However, most of his laws hold up without any amendments. Most modifications concerned phenomena that Newton had not studied, such as extremely high velocities. No, only new phenomena for which no laws have yet been identified might lead to the discovery of new regularities. Only those might bare the promise of new applications. Such unexplored territory was much more commonplace in the 19th century.

However, the above discourse is likely to end up in the same place as the prophecies of the 19th century scientists; the bin. So be it. You've read my book this far. But why shouldn't I, as a physicist, be allowed to fantasize about what *is* possible, about avenues that have not yet been explored and about technological developments whose limits are not yet in sight? What day dreams can we allow, if we wish to abide to the rules, and obey the laws of nature? Physics hasn't 'finished' quite yet; nanotechnology is only just taking off, there are lots of potential space projects to be imagined and communication via computers is only a few decades old. There is lots of room for expansion here. Let's see how far we can take it.

In the next chapters I will tell you what you may and may not expect of space travel, what the information technology revolution still has to offer, which serious changes to our society are to be expected, and which ones aren't. Every now and then the dry language of physics might put you off, but most of what I have to tell you can easily be followed with a bit of common sense. I hope to surprise you. Even within the boundaries of real physics, we can make dream worlds come true; worlds where the laws of physics continue to keep everything under control.

Not everything I will tell you about are true prophesies. There will often be reasons having little to do with technology or physics that will prevent the use of certain inventions. Can magnificent constructions of the future be sufficiently safeguarded against terrorist attacks, for example? I will not take these aspects into consideration, when I talk about future possibilities. Sometimes there will be economic, political or ethical objections against certain developments, such as the transfer of biological life from Earth to another planet, or against some of the more fantastic and fascinating possibilities that I will disclose later on.

There will be small details that will disappoint you. Such as my conviction that the exploration of extra-solar planetary systems will take tens of thousands of years to accomplish. This means that neither you nor I will be there to admire the results of such expeditions. Nevertheless, there are many other fun things to look forward to in our own time.

Exploring new developments and new ideas, and also repudiating those that are incompatible with our understanding of the laws of physics, that's what this book is about. Actually, I should have molded what follows into a juicy and romantic science fiction story, with good guys and bad guys, with a plot and lots of sex — after an unbelievably narrow escape, the good guy forges a miraculous happy ending — or something to that effect. But this type of creative writing is not my forte and it would distract too much from what I really want to talk about. You'll have to use your own imagination as to how the would-be heroines of these stories boldly go where no man has gone before....

chapter 2

Take Off

PRESIDENT KENNEDY HAD made a promise, and he had kept his word. Within a decade, an American had set foot on the Moon. It was July 1969. I was glued to the television, mesmerized. Interplanetary travel was now within our reach. At that very moment, space travel was proven beyond any reasonable doubt. The biggest obstacle was getting off the ground. We have all seen it on television; apparently, blasting off into space requires an immense firework display.

Now this happens to be hard science. To get into an orbit around the Earth, a spaceship has to reach a velocity of about 7.5 kilometers per second (27,000 kilometers per hour) very rapidly. A rocket engine would be able to reach such a velocity most efficiently, if it could generate a gas stream with two-thirds of that velocity. In other words, a gas stream with a velocity of five kilometers per second. Alas, the laws of chemistry do not normally allow for such velocities. Chemical reactions can only generate a velocity for gas molecules of up to four kilometers per second.[a]

---

[a] There is the possibility that future technologies will allow the use of such exotic chemicals as mono-atomic hydrogen. Such fuels could generate velocities well in excess of five kilometers per second, but at this time, no one knows how to safely store such fuels.

The laws of physics then determine the velocity that can be reached if the total weight of the spaceship is halved by burning fuel; its velocity will not be increased by much more than about two kilometers per second. Empty fuel tanks are discarded, and this process is repeated. The weight of the spacecraft is again cut in half, at which point the velocity is doubled. Once the Earth's orbit has been reached, the remaining weight of the spacecraft, which we call 'payload mass', is only a small fraction of the weight at take-off. That is why most launches into orbit are conducted in phases, such as with triple-stage rockets. In any case, we all know that the initial mass at take-off is significantly larger than the payload mass that enters into the Earth's orbit.

Well then, can't this law of physics be circumvented? Aren't there ways to obtain such velocities by other means? Perhaps by using nuclear fuel. Nuclear fuel produces much more energy and could generate a much faster outflow of gases. However, splitting atoms creates almost a million times more energy, which causes a whole different set of problems. To take off from Earth, a force larger than Earth's gravity has to be generated; that is an acceleration of ten meters per second per second, or 10 m/sec^2. This means the outward flowing gases must be sufficiently dense. However, if the velocity of the gas flow was much more than five kilometers per second the amount of energy created would be enormous. This would pose an irresolvable cooling problem.

Is it possible at all to attach to a spacecraft a nuclear reactor, which transfers its energy to the gas flow efficiently? If the question is to get a spacecraft off the surface of the earth, it is not entirely unthinkable, but modern methods using chemical engines appear to be the most elegant and practical. In practice, it has proven to be the most effective. Let me add this: launching a spacecraft and creating high velocities in space will always require a lot of energy, whatever technique is used, simply because a spacecraft requires high amounts of energy to generate movement and to lift it all the way from ground level to a stable orbit in space. It is possible to calculate how efficiently the available energy is used in various acceleration techniques, and usually it is pointed out that rocket engines waste a lot of energy. It certainly doesn't look very efficient, to push yourself up by blasting large volumes of gas downwards. However, calculations show waste to be minimal. The most efficient type of rocket would be one that is able to control the velocity of the gases emitted. It would start by emitting gases at a low velocity, increasing that velocity as the voyage continues. One can calculate that such a rocket could be close to 100% efficient: *all* of the energy released in its exhaust gases would be turned into kinetic energy of its payload mass. If a spacecraft is launched into orbit using an engine that has a fixed exhaust velocity of three kilometers per second, fifty percent of this energy is lost. This is not at all bad, considering the lack of alternatives. Other possibilities to travel through space will be examined later: are there alternative ways to come off the ground? And incidentally, how far can a spaceship travel altogether?

As an aside, you must have noticed that I am using kilometers instead of miles. I will also talk of kilograms and centimeters. These are the metric units, the units of science. Regrettably, much of the Anglo-Saxon world continues to exploit Imperial units: miles, inches, pounds, ounces and a whole load of mutually incomparable, ancient concepts. Maybe you prefer these when you drive your car or do your shopping, but in science the metric units really are a lot more convenient. On 23 September 1999, an unmanned space vehicle, the Mars Climate Orbiter, arrived at the planet Mars. Then something went wrong. Radio contact was lost, and soon it became known that the $327.6m apparatus had crashed upon Mars. An investigation afterwards identified the cause of the failure: some piece of computer software back on Earth was still using pounds as a unit of force, while the spacecraft was tuned to Newtons, the metric unit. The Mars Climate Orbiter was intended to enter orbit at an altitude of 140 to 150 km above Mars. However, this navigation error caused the spacecraft to reach as low as 57 km. The spacecraft was destroyed by atmospheric stresses and friction at this low altitude.

I remember the disappointment. This would have been a beautiful observatory near Mars. I felt that the guy who had still been using pounds and ounces should have been on board that spacecraft! NASA's final report explaining all the factors that led to that disaster listed hundreds of contributing reasons but neglected to mention the main reason — that the USA has failed to go 'metric'! Anyway, I am using metric units throughout the book. Just in case you didn't know, one kilometer is 0.621 mile, one meter is 3.28 feet, and a kilogram is 2.205 pounds. A Newton, by the way, is the force needed to accelerate one kilogram by 1 meter per second-squared,[b] or 0.225 pound-force.

---

[b]That is, the velocity increases by 1 meter/second every second.

chapter 3

Inside

AFTER YEARS OF research, Professor Prtplwyszpo finally accomplished what he was after; he had constructed a shrinking machine. Whoever ventured into the machine through the little trap door emerged from the other side roughly 10% smaller. If the volunteer repeated this process seven times, he would shrink to less than half his original size. If passed through the machine often enough, the original size could be reduced to whatever fraction you wanted. After a hundred cycles, a medical team would be able to enter a patient through his nostrils and travel to the affected area in the body to perform revolutionary operations with microscopic instruments. After successful completion, the team would walk back through the machine exactly the same number of times to regain its original size.

That would be so neat. But what would happen to the atoms and molecules of the medical team's bodies? Would they shrink too or would a few of them be removed? In the latter case, the nuclei of the cells in their bodies would be damaged beyond repair and the individuals would die instantly. Every cell in our body contains molecules called DNA, which control the functions of those cells. Every atom of the DNA molecule is of vital importance; if you remove 10%, you'll die. The reader can probably identify other erroneous assumptions in this science fiction story; it was a nice idea, but it simply can't be done. What actually *do* we know of the world of little things?

It sounds reasonable; to discover the world of the minuscule, you would need very tiny instruments. Can insects see other insects better than we can? With their small eyes, couldn't they get much closer? Well, if you thought this to be the case, think again. Our eyes are very sensitive instruments that observe the world with the aid of light. A fundamental law dictates that the bigger the detection machine, the better it works. It is actually much easier to uncover the secrets of little things with bigger eyes. Our eyes are bigger than those of insects, so we can see insects much better than they can see each other or their own world. A fly keeps flying into the window because it can't see the dust or the reflection in the glass. Similarly, whales have much better hearing than we do. So when it comes to visual and auditory sensory functions, size matters. It is therefore not surprising that some animals have evolved to become as large as possible.

Thus, to detect the tiniest particles of matter, we need gigantically large instruments. A laboratory complex named CERN, located near the city of Geneva in Switzerland, is a collaborative effort of a large number of European nations. The abbreviation stands for Conseil Européen de Recherche Nucléaire. More accurate would have been to call it sub-nucléaire; the particles that are being studied are becoming smaller and smaller. Within the complex is a large ring-shaped tunnel, partly on Swiss soil, partly on French soil, with a total length of more than 25 kilometers. Within this tunnel, particles are accelerated to great velocities in opposite directions. When the particles collide, the effects are registered with instruments varying in size between a truck and a house. These effects are registered millions of times per second. This has proven to be the most successful way to analyze the properties of matter on an extremely tiny scale.

This kind of research is called high-energy physics, because it requires pumping the largest possible amounts of energy into the particles to reveal the tiniest details. This is because the particles must reach a very high velocity before we can accurately measure their positions. This principle is directly related to Heisenberg's Uncertainty Principle, to which I referred in Chapter 1, which states that it is not possible to simultaneously and precisely identify a particle's velocity and its position. Scientific theory now dictates that investigating small distances requires a range of high velocities.

There are several laboratories in the world, like CERN, where similar research is being conducted, namely in the United States, Germany and Japan. Their studies have resulted in spectacular new insights into the composition of matter.

All matter that surrounds us, including ourselves, consists of particles we call molecules. Every molecule consists of units called atoms. An atom consists of a very small nucleus, circled by electrons. Under normal circumstances, atoms are robust unchangeable marbles. However, atoms can rearrange themselves to form different molecules; the outer electrons dictate how atoms get attached to or detached from one another. Chemistry is based on this principle. This rearranging may generate energy, usually in the form of heat, such as with combustion. An atom may also absorb energy, for example in the process of photosynthesis, when plants use sunlight to create organic molecules from air, water and minerals. But let's continue.

As far as we have been able to determine, electrons are so infinitely small they appear to be mere 'points', in the mathematical sense of the word. Actually, this simply means that we have not been able to identify their spatial structure. Space in their immediate surrounding is a little distorted or polarized, and this distortion can be calculated quite precisely, so electrons do have structure in this sense.

An atomic nucleus, on the other hand, has a lot more internal structure because it is made up of two types of particles: protons and neutrons, and they are held together by other particles, named pions. Protons and neutrons can only be rearranged across different nuclei through a nuclear reaction. This involves a lot more energy than the amounts seen in chemical reactions; sometimes a million times more. This is what we call nuclear energy. Only through nuclear reactions is it possible to transform one atom into another.

The discovery of nuclear reactions, with the enormous amounts of energy that they may release, was one of the most important breakthroughs ever made in physics, having the most dramatic consequences for society. Because of the inherent dangers, nuclear physics has met with severe opposition from the public. However, nuclear physics can be used constructively to address societal problems. More about this in Chapter 8.

Protons and neutrons are also not immutable. Every proton and neutron contains three quarks, held together by particles we call gluons. Pions, which hold together protons and neutrons, consist of a quark and an anti-quark. Quarks and gluons are, just like electrons, 'points'. Further, there are all sorts of exotic quarks and exotic variations of electrons, but these are rare and only have a brief existence. The energy required to reposition quarks is even greater than that needed for nuclear reactions.

There are other types of matter particles, such as neutrinos. They are extremely inert, have little mass and are very difficult to detect, but with our gigantic instruments we have accomplished it. Then we also have what is called dark matter, of which we know next to nothing, except that its presence is given away by the gravitational forces that set visible matter in motion. Dark matter consists of particles that must be distinctly different from those we know of today.

We have come to the end of my summary report of the world of the small. But why am I telling you all this? Well, I am often asked what we could actually do with this bit of science. Could we use those quarks and electrons to generate energy? Does a quark computer exist? Science fiction stories live off these ideas. Reality, unfortunately, is more sobering. It is conceivable that totally new discoveries will be made, such as a new class of elementary particles that could serve as catalysts for nuclear reactions, which are currently thought to be impossible, and which would generate even more energy than what can presently be achieved using nuclear reactions. Physics as we know it tells us that this is possible in principle; the so-called 'magnetic monopole' could be such a particle. Talking about this particle goes beyond the confines of this book, but with the aid of these particles all matter could be transformed into energy, and not just a small fraction, as with currently known nuclear reactions. Unfortunately, we have no concrete evidence that such a particle exists, let alone that we will ever be able to create one. It is highly likely that this will never be possible, but who knows?

The characteristics of elementary particles can only be identified when they are smashed against one another with a lot of force. A lot of energy is required and usually a lot of energy is lost. Therefore, the instruments we use for studying quarks and gluons also absorb large amounts of energy. That is not really a problem if all you are interested in is studying their characteristics; on the contrary, those particles pose a beautiful challenge. But it doesn't render them very useful, and practical applications, such as a quark computer, are not easily imaginable. Rather, we should compare this topic with our study of faraway stars and galaxies; important, essential even, for our general understanding of nature and our place within it, but in practice out of reach for colonization. More about that later.

What then is realistic? Could we make tiny appliances? Computer elements? Robots? Detectors? And how small exactly? In practice, the limit will be at the atomic level. Atoms, with their complicated chemical interactions, can do the work for us. They are already doing that in every chemical factory. The most fascinating aspect, however, is when you realize how many atoms exist and how complicated their interactions are when you put them together. And if you then realize how small these atoms are, it isn't difficult to imagine how much more there is left to discover. We do not need quarks; it is more realistic to project our reveries about the future onto those atoms.

The complicated nature of the world of atoms, what they can be used for and the various complicated chemical interactions they can be involved in, are evident in everyday life. All life is based on it. All information regarding the composition and functioning of any living organ is stored in thread-shaped DNA molecules, and body cells can access that information instantly, as if they were advanced supercomputers. Sciences, such as biochemistry, have only scratched the surface when it comes to recording all the information that is stored in DNA molecules, and also when unraveling how the computer in each living cell processes that knowledge.

Ideally, we want to be able to analyze objects and to construct or reconstruct them atom by atom. We are as yet nowhere near being able to do such things. However, I am much more careful with my veto on this issue; there is still a long way to go and we will certainly hear a lot more about this topic.

chapter 4

Computers

It was the early '80s. As a physicist I was connected to a computer network, which I could use to send messages to all my colleagues in the whole world. Sometimes it would take only half an hour to receive a reply! What a privilege, and how practical! It combined the advantages of a telephone call and a written letter: the messages were delivered with the speed of a telephone call, but with 'email' one could send detailed and precise messages without having to disturb the recipient; your mail would be read at whatever time was convenient. If only other folks knew about this! But at that time, no one besides physicists had heard of 'email'.

This sure has changed. Our society has been in the iron grip of a significant development for decades: the information revolution. Not so long ago, telephone, radio and television brought about revolutionary turning points in communications; now it is the personal computer. The general public has discovered the comforts of having a PC. Between 1998 and 2002, the number of Internet sites grew from 3 million to 900 million, and the computing power of our PCs has increased tremendously. This is because the number of transistors on every single chip doubles every 18 months. This trend had already been noticed by one of the founders of Intel, Gordon Moore. Moore's Law, as the trend has been dubbed, continues to be valid today, albeit that the doubling period is usually said to be closer to two years than 18 months.

And this is only the beginning of the information revolution; there is lots more to be expected from it. As we become able to make very small and cheap electronic components, the transmission and storage of information will become even more efficient, a change that will reverberate everywhere.

Take photography, for example. Only a decade ago, everyone was still using photographic film and it took a week or so before you could see your shots. Now, not only do cameras give you instant digital images, you can make pictures and even videos with your cell phone. This is due to the enormous increase in the capacity of single chip cards. Video cameras were bulky devices twenty years ago, but now they fit in your mobile phone, which itself is not much bigger than a credit card. Vehicles contain more and more computer-guided devices. Toys contain electronic gadgets and computer games are becoming fancier with every generation, and these generations follow one another incredibly quickly. Daily news and other information reaches us in an instant. TV images no longer require a fixed television set at home, but can be received on your telephone as well.

We see all of this happening today and much of it had been difficult to predict for most of us. What will come next? Sensors for motion, sound, temperature and smell will become tiny and cheap. What will they be used for? Homeowners might like to turn their house into a living being, which not only knows whether someone is at home, but is also informed in advance when someone arrives, so that it can automatically control the temperature, switch the lights on and off, even have the coffee ready when you arrive. Of course, the house senses if an unauthorized person enters and knows when to alert the police. All valuables in the house, by the way, are equipped with chips as well, notifying the owner of their whereabouts. Simple gadgets like toothbrushes, kitchen tools and so on will be equipped with chips containing instructions for their use; they will do most of what you want without even you asking.

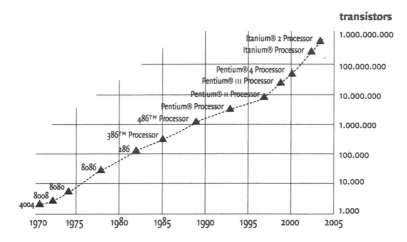

Moore's law, during the last 35 years. On average, every 21.5 months the number of memory elements on a chip doubled.

Where will all this lead to and where will it end? The limits are not yet anywhere in sight; the world will undergo even more drastic changes. While I am writing this, most computers have a memory capacity of hundreds of megabytes located on chips of just a few square centimeters. There are dozens of elementary memory cells per square micron on the surface of a chip. A micron is a measuring unit: one-millionth of a meter or one-thousandth of a millimeter. The limitation on the number of memory cells per square micron is presently dictated by the wavelength of light, as I will now explain.

All the chips in your computer are made of semi-conducting material, such as silicon. Semi-conductors are substances that, when they are very pure, conduct electricity, but only poorly. If atoms of a different type are added to the pure material, then a slight mismatch may arise between the number of electrons in the material and the positions they are expected to occupy between the atoms. Thus, 'superfluous' electrons are created that are only loosely attached to the atoms, and they conduct electricity very well.

If other types of atoms are added, then empty spaces are created where electrons should have been. These empty spaces are very mobile, just like free electrons, and these substances will therefore also conduct electricity much better. By positioning electrons against empty spaces, the material can be manipulated, which enables us to produce regions with different kinds of electronic properties. One can produce electronic devices by giving these regions the shape of very complicated microscopic racing tracks. How is this done in practice?

The trick is to use a microscope in its reverse: extraordinarily clear images of the desired patterns are projected onto slices of silicon or some other semi-conducting material. Then, chemical substances are used that are sensitive to light. This is called 'etching'. Through etching techniques, memory elements are formed with wiring and all. The challenge consists of making these as small and efficient as possible. The imaging procedure was performed using light. Therefore, the limiting factor at present is the wavelength of the light beam that is used to do the etching.

Physicists have already discovered the advantages of ultraviolet light, which has a much shorter wavelength than visible light: only a tenth of a micron. But that appears to be the limit of this optical method. If the wavelengths are too short, there is no material to make the optical lenses that focus the light rays. Sooner or later, the need will be felt to go to even smaller sizes, which would enable one to reach even larger memory capacities and greater processing speeds; but then the optical method will have to be abandoned. Could we take it any further? Can we go any smaller?

In December 1959, the American theoretical physicist Richard P. Feynman gave a lecture to the American Physical Society at the California Institute of Technology. The title of his lecture was 'There's Plenty of Room at the Bottom'. Feynman suspected even then that it must be possible to store enormous amounts of information if you could produce memory cells measuring just a few atoms across. He fantasized further about what could be accomplished if it were possible to make objects as small as you wanted, and if you could control matter atom by atom. During his lecture, he was looking much further ahead than his contemporaries.

The current computer chips consist of elements that are already very small, but not quite as small as Feynman had imagined.

We humans rely on light to look at things; eyes, just like lenses, are useful to observe large items, but not for microscopic and submicroscopic objects. Insects prove our point because they have very small eyes and do not see well. Instead, they have additional sensory devices, such as antennae, which they use to examine their surroundings. Even smaller creatures have bristles and quills, and an exceptional sensitivity for chemical substances, but no eyes or if they do, very bad ones.

Scientists have also discovered this. Microscopes based on the sense of touch already exist, known as 'tunnel microscopes'. A very sharp needle approaches the object to be studied, and an electric circuit 'probes' to make sure it does not get too close by detecting whether electrons jump from the needle to the object or vice versa. This is known as the 'tunnel effect'; the so-called 'potential barrier' between the needle and the object is still too high for the electron to pass through, but the laws of quantum mechanics allow it nonetheless, as if the electron digs a little tunnel through the potential barrier. The strength of the electric current that is created by this process depends on the distance between the needle and the object: the smaller, the stronger. With this technique, it is possible to create fantastically clear images, using the same method as tiny insects.

Even individual atoms can be visualized this way. The same needle can be used to pick up atoms from a surface and move them around. In this way, it would be possible to create things on an atomic scale, including computer chips. The sharpness of the needle is the limit, but the tip could also have atomic dimensions, that is a thousand times smaller than optical images. This technique is already mastered in laboratories. As early as 1990, researchers at IBM succeeded in arranging atoms of the element xenon in the form of the company's logo. Why then are computer chips not yet fabricated in this manner?

Unfortunately, there is a problem. Of course there would be, otherwise these methods would already have been applied on a large scale. The probing described above takes time; a lot more time than is needed to produce optical images — those you can create all at once, while the probing method requires a point by point approach. In the chip manufacturing industry, time is money. Time is of such importance, that it makes economic sense to acquire all kinds of extremely expensive optical machinery. I've been told about a new machine comprising enormous quartz lenses, which are used to treat an entire 30 cm diameter slice of silicon with sharp ultraviolet light projections. The old machine, which could only handle slices with a diameter of 20 cm, was discarded because the newer machine enabled the production of a larger number of chips per minute. And that's what counts most for those manufacturers. So forget the probing method. It would produce faster computer chips, but I'm sure they'd be way too costly.

But can't we think of a faster method based on probing? There must be one, I think. The way I imagine it, the central element of such a chip assembly machine is a comb with teeth of atomic scale. Every individual tooth is controlled electronically. Materials of every desired type would be spread on a piece of silicon crystal, like butter on a piece of toast. The optical fabrication of a chip currently takes an average of no less than ten seconds; my comb could do this just as quickly. Perhaps the comb would be made with lots of holes, instead of teeth: ultraviolet light, or beams of atoms or electrons are sent through these holes, which are opened and closed electronically. I don't know how realistic this fabrication method is, but what I wanted to show here is that the theoretical limit for economical and usable chips is much further removed from where we are at present, and that's why it makes sense to suspect that the computers of the future will be much faster than the present ones and will have much more capacity. Moore's Law will continue to be valid, at least for the time being. The optical method is sure to be abandoned, but finding an alternative will take extra time. This may be the explanation for a complaint I hear regularly, which is that the computers of late have not become that much faster or bigger than their predecessors, as if Moore's Law is no longer valid.

Maybe even more innovative ways will be invented to store information. Thinking back to Feynman's ideas, maybe we can create microscopically small hard disks, thousands of microscopic disks on a slice of silicon. The information is added and read by microscopic needles of atomic precision. The little needles move individual atoms or molecules around on disks, the same way you would draw figures in a sand box. More than one memory element per atom is not feasible, but we will be able to approach this limit. The practical limit, then, is moved forward to a few hundred million bits per square micron, or tens of billions (10 to the 13th power) of memory elements on one chip. Whether my miniature hard disk can really be manufactured, I don't know. There could be fundamental obstacles involved in working with moving parts of such small dimensions, such as instability, wear and tear, and friction, but I think the atomic scale boundary is feasible in principle.

What we lack is experience. Humanity has only just begun to explore these types of concepts. However, I have complete faith in human ingenuity and there will be numerous clever findings that will enable us to achieve the most from every opportunity we discover. And unlike 1959, there are many laboratories around the world investigating nanophysics today and there are many researchers who are dedicating their lives to this. The science that deals with the manipulation of individual atoms, or small groups of atoms, is called nanotechnology. A nanometer is one-thousandth of a micron. Ten hydrogen atoms fit onto one nanometer.

A lot is to be expected from nanotechnology. Let me give you an example. The hard discs of present day computers are equipped with what looks like a gramophone record, which stores all the information that needs to be saved when the computer is turned off. I believe that such a hard disk will disappear in the foreseeable future. We will switch to using microscopic memory elements that retain the information also when the power is switched off. These could be microscopic hard disks, such as those described above, or completely different devices. Starting up computers and specific programs will take considerably less time. Computers will therefore be much more energy efficient. Permanent memory elements are already used in cell phones and the like, so general usage in computers is only a matter of time.

The illustration below depicts what we think machine components of atomic dimensions will look like. If you compare these to our current tools, the latter look like products straight from the Stone Age! We could think of all kinds of applications, such as sensors and measuring instruments that don't take up any space, weigh nothing and are incredibly quick. We could adorn the moving components of all kinds of tools with these types of sensors, so that any malfunctions can be traced instantly. Alarm systems could be much more specific; they would be able to establish whether there really is cause for alarm and report in much more detail what is actually going on.

Instruments of atomic dimensions could be used for any purpose imaginable, but particularly in medical sciences. Remote-controlled robots could be sent to any part of the human body to resolve

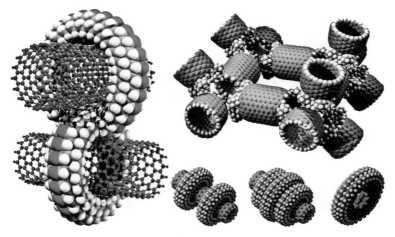

Several nanostructures, connectors and layers of atomic dimensions.

problems. Indeed, in this manner we could still realize the dream of the shrinking machine but it is not us who are shrunk, but rather our robots. Less advanced, passive robots could be attached to capsules containing medication; the capsules would be injected into the body, with the robots ensuring administration of the medication in the right dosages for weeks or even years. (Much more about robots and remote control in Chapters 6, 12 and 13.)

It is only a matter of time before very small computers become commonplace. Thinking machines that are only centimeters or millimeters big, which, because they are mass produced, will be very cheap. They will be used for everything: domestic appliances, cars, toothbrushes, bicycles and all kinds of objects that must be protected from theft.

Moore's Law, remember, prescribes exponential growth. We know from experience that such growth will not last forever — it will end, but when and where? Could the elements of a chip really become as small as a group of atoms? We will probably encounter a few fundamental problems along the way. Every operation performed by our computer element costs energy, and this in turn generates heat. This heat has to be disposed of immediately, otherwise the chip will melt. Random motion caused by heat can become troublesome, particularly at very small scales. That means that if the shift between a 0 and 1 or vice versa requires too little energy, heat fluctuations could generate such a shift spontaneously. This would lead to a computer's malfunctioning. We don't want that, so that will be a limitation.

There was something else in Feynman's speech. He thought it would be possible to make a three dimensional box of information units, rather than spreading it out on a surface, like with our current computer chips. If you could store one bit of information in a little box made up of 100 by 100 by 100 atoms, then all knowledge gathered by humanity to date would fit into a little parcel of matter not much bigger than a dust speck. The ultimate limitation to the memory capacity of computers is therefore dictated by the number of atoms per volume unit. That it is possible to make the storage space for information very small has already been proven; living organisms contain complete dictionaries of all inheritable features of their bodies in the form of DNA molecules, and one complete dictionary takes up no more than a fraction of each living cell.

In other words, electronics has not yet reached its theoretical limit, not by a long shot. Memory elements have the potential to hold millions more bits of information than they can at present. The speed of calculators could also be substantially increased. The heat that is produced is already a significant problem, but there is also ample room for improvement.

Strictly speaking, the theoretical boundary of electronics is not sharply defined — because there is something else. At the atomic level, the processes follow the dynamic rules of quantum mechanics. This results in various complications, for example, electrons move more erratically, passing through barriers they are not supposed to. On the other hand, quantum mechanics is very special and maybe we will be able to profit from these complications. The so-called 'quantum computer' is a theoretically conceivable construction, in which an unlimited number of calculations can be carried out simultaneously by one and the same computer element. That such a device is feasible can be illustrated without complicated technical arguments. Quantum mechanical phenomena can create situations that are extremely difficult to calculate through the use of equations. This situation might be exploited to our benefit, by reversing it. With the use of a quantum computer, we create an arrangement which was identified as the initial state of such a complex 'quantum process'. We know the equations for this process and we choose the arrangement in such a way that the equation corresponds with the complex calculation we want to conduct. Nature will carry out the desired calculation for us posthaste, with a minimum of energy dissipation.

We have not yet been able to construct quantum computers, even though there are supporters who claim that the theoretical principle has already been demonstrated. But it is questionable whether the practical obstacles can be overcome. And even if that can be accomplished, then the types of calculations that can actually be carried out by a quantum computer are only very limited. A well-known example of what such a computer would be able to do is breaking secret codes, because what a quantum computer will be very good at is searching for that one combination of numbers that is the solution of a very complicated equation.

The future 'brains' of our computers would not have to be any bigger than the tip of a pin, provided we would not expect a lot more intelligence from our computers than we do now. But we will. Our computers will also be a lot more powerful. I'm afraid, though, that the software manufacturers will come up with all kinds of useless functions that will absorb all the extra brain power, so that the starting up and closing down of our computers will yet again consume unnecessary lengths of our precious time. After all, that's how it has been in the past.

Information of all sorts is becoming cheaper. We are already seeing the trend of adding a timer to pretty much every consumer household product — time is very easy to measure nowadays, and accurately as well, so why not? Our cars and houses will be decked out with detectors, which will make all kinds of worries redundant — Did I leave the light on? Are my brake lights functioning? Is it time to service the car? Can I reverse another ten centimeters? Has the thermostat been set to the right temperatures? Who is entering my house? GPS systems are currently considered a luxury, but these, and many other devices, will be common in due time. Not because we really need them, but because they are so cheap — and we won't have to sit in slow traffic if we have a detector in our cars that suggests alternative and faster routes. Those who do not yet have an infrared light sensor in their cars now pose a danger to other travelers, because most others will use them already and no longer have any reason to slow down in foggy conditions.

Digital photography, another example of such a new development, has achieved its bid for dominance. Rolls of film that still need to be developed the old-fashioned way are left to rot on the shelf. Printing pictures is also not really necessary anymore; we can look at them on our computer screens.

Even those in space travel had not seen this development coming. Richard Greenberg writes about this in his book, *Europa, the Ocean Moon*. When the space probe Galileo took off to Jupiter in 1989, it was seven years behind schedule, and the technology for producing pictures by the probe was already outdated in 1982. It was decided then not to send more than 100,000 pictures back to Earth. The reason being, that the brothers Hunt were trying to buy up all the silver deposits in the world to attain a monopoly of the precious metal, so the high price of silver had to be taken into account; at least several copies of each photo would have to be developed to make them available to scientific researchers and the high price of silver would cause serious budget overruns, according to contemporary expectations.

Well, by the time the space probe arrived at Jupiter in 1995, the pictures were sent to the researchers by email and viewed on their computer screens. The scientists regretfully admitted that they had not foreseen that optics and electronics would go so well together. More about that, later.

Moore's Law is an exponential law, so it will undoubtedly come to an end. I suspect that its implementation will slow down over time, and that the doubling period will be four years instead of two, and then eight years, and so on. The need for memory space will probably also lessen over time. I am curious as to how society will react to the fact that our products can no longer be improved, but I don't know when that will be. If I look at what we might be able to achieve at atomic level, Moore's Law will probably last for another 60 years or so.

In the meantime, communication is also becoming cheaper. I was surprised by the quick advance of cellular phones and I was just as surprised by the fact that this has not led to greater technological problems. Sometimes you see dozens of people making phone calls in the same place, at the same time, but apparently the bandwidth is large enough — you never really experience any jams. Cable companies are falling over each other to offer more and more television channels; glass fibers are able to process enormous data streams, but also more and more information goes through wireless connections. The ugly clots of cables on and around our desks are fast becoming obsolete.

Physicists are working on something new with regard to the Internet. After the World Wide Web, there will be a World Wide Grid, a network of computer connections that can transmit large files incredibly fast. This way, parts of large scientific research projects can be outsourced to various researchers across the world. Possible examples are large scale weather predictions, but there are other scientific fields that require transmisson of enormous amounts of data, such as astronomy or particle physics. And there are also more everyday applications. Hospitals, for example, often make MRI scans of their patients. The 'WWG' would be able to send those large data files to other hospitals quickly and frequently.

chapter 5

Paper

A LARGE PART of the information revolution is still ahead of us, and exactly what it will bring is mostly subject to speculation. Many products will be enhanced with much more elaborate control systems than we have now, so that tools will be cleaner, safer and more efficient. A good example is television and computer screens.

Nowadays, anyone who writes a letter, reads a book, newspaper or a magazine, uses paper. Printing is so easy that more paper is being used than ever before. But why don't we just read off our display screens? There are quite a few reasons for that. First of all, the software programs that have been thrust upon us are abominable. While I'm writing this, I'm fighting with an unruly text editor, whose help files are anything but helpful; they are extraordinarily long, unreadable and unusable. It is not possible to solve a problem without much ado, quite the contrary: its creator has added all sorts of features to the program, but which way they are supposed to be used is completely unintelligible, so I just don't bother. And that is only the tip of one great iceberg. This indecipherable user-unfriendliness is found in almost all software. There is so much more work to be done here.

Let me start by giving the software manufacturers one tip: never ever let the designer of the program write the user manual, the

FAQs or the help texts. They never realize the sheer magnitude or the nature of the problems the innocent user has to put up with, let alone the fact that the user does not want to memorize 30 pages of information before being able to solve a simple problem. How often does the box of your brand new product say: *"Dear user, please read the manual in its entirety before using this appliance"*? Of course, we reply: *"Dear manufacturer, that's more than 30 pages, I don't think so."* And surprise surprise, the first problems instantly arise. But I don't really have to tell you this, as I am sure you have experienced it yourself. Well, in the not-so-distant future, I expect that appliances will not require any written manual at all. You will just flip the switch and if you do anything wrong or you don't understand how it works, a short and clear piece of text or a friendly voice will assist you — or at least this is what I hope for in my more optimistic moments, though I can't be sure that the manufacturers will want to put any effort into the creation of such facilities. But I am certain that, technically, it can be done easily.

Anyway. We asked ourselves why we continue to use so much paper. Well, our current display screens are awkward, bulky, slow and unfocused, so you are not really going to take your computer to bed to have a quick read before going to sleep. Flipping through the pages of a magazine, circling something of interest, filling out a crossword puzzle or tearing out an advert on screen is also not possible. Furthermore, the display screen is often too bright or too dim, depending on the ambient light of the room.

Now, imagine a screen that is not only flat, but also lighter than a book, has clearer imaging and a built-in illumination feature that automatically adjusts itself, assimilating to your surroundings instantly. Moreover, it is easier to use than a book because, for instance, it will be much easier to retrieve that page you were looking for. It is certainly imaginable that paper will become redundant. Not because we want to get rid of all paper, but because we have stumbled upon a more convenient medium.

To date the situation is quite the opposite: even though the display screens are improving, more paper is being printed and copied than ever before. Paper is pleasant, familiar and easy, but paper is also a type of medium that can be written on only once and then has to be discarded. How expensive and wasteful.

Maybe there will be a time when we consider paper a mere inconvenience. Then, we wouldn't buy a newspaper anymore, but would download the news off the cable and read books straight off the Internet, because why would we want to buy such a heavy thing that takes up lots of space and destroys our forests? This will happen only if downloading a book is made easier and more enjoyable than browsing the local book store, and if the newspaper that is downloaded off the Net every morning onto your mobile display screen is easier to read than pages made of old-fashioned paper. At the time of writing, nothing is further from the truth.

Yet display screens are invading our homes and offices more and more. Large and colorful displays are becoming cheap, efficient and ubiquitous. Tiny displays will adorn our kitchen utensils. Our cars will be full of them. Can you imagine a world without such screens anywhere?

Actually, I can. Perhaps conventional displays, in turn, will be replaced by something even more convenient, more economic and less costly: for example, high-tech 'viewing glasses', which are connected to our computers, handy for those who want to read in bed, very light to carry, and energy efficient. But again the same caveat: this will only happen if the spectacles are designed to be more user-friendly than both a display screen and printed paper. How would the glasses work, you ask? In fact, early versions are already being made today. Some of these resemble small, lightweight opera glasses with individual lenses, focusing on tiny screens, one for each eye. In the future, the screens might be controlled individually, so that it would be possible to create the most wonderful 3D effects. They would still require the use of a mouse, a keyboard or the likes to identify what you want to look at, but perhaps your voice would suffice. Head movements would be recognized, to enable a fantastic feeling of virtual reality. Maybe at a later stage it would be possible to produce even more advanced glasses, which monitor eye movements, so that the glasses can be made even smaller and lighter, and perhaps even fit as comfortably as contact lenses.

In the future, we expect life-changing progress in the medical sciences, particularly in relation to our eyes. Despite the expected improvements, eyesight deteriorates when we age, and with demographic aging this may well continue to be a considerable problem. Can we produce appliances that will assist the elderly with their reading?

Indeed, my mother-in-law received a fantastic aid recently — a book is placed on her machine, which scans the pages, recognizes the text and then reads it to you in a pleasant tone of voice. It is a great improvement compared to what existed only a few years ago, but the first of such machines was already constructed in 1976! The machine works wonderfully: it recognizes text that is written in columns, and it recognizes advertisements and medication instruction leaflets. It even manages to read coloured letters against a coloured background! But, my mother-in-law tells me, if the text includes a word in a foreign language, then the machine is stumped and she has to ask it to spell it. And a handwritten text is entirely beyond its command.

"Excuse me? What's that?" is my response; they can't include a tiny program in all that amazingly beautiful software that understands other languages? And my mother-in-law, however happy she is with the machine, still needs to turn the pages herself. And this is my reaction to many modern appliances: I'm delighted that someone managed to invent it, but surely it can be much improved upon? In no time, my mother-in-law, at least that's how I envision it, will toss the *Woman's Weekly* towards her robot, who picks it up and reads it to her, translates difficult words and tells her what's in the pictures. Maybe it will even recognize movie stars and celebrities.

And by the way, the *Woman's Weekly* will soon be available only in electronic form. "Too bad that robot of yours learned to pick up magazines and read them for nothing," one of my colleagues joked.

In any case, we want robots to be able to do a lot more, not less. Reading the *Woman's Weekly* was only entry level robotics.

In relation to handwritten texts, I know that the current machines have tremendous trouble reading them, but their abilities slowly improve and I think that the problem will be solved some time soon. It is only a matter of time before machines understand handwriting better than we can, because they have access to larger databases and they will possess the knowledge as to how brainwaves result in hand movements. In the same way, they will be able to recognize heavily distorted letters, guess the missing parts, and so on.

The most revolutionary application of the increase in information will undoubtedly lie in the area of artificial intelligence. Science fiction has educated us in a somewhat biased manner — it makes little sense to model robots after human form. Intelligent robots won't have to carry their brain on their shoulders — it is a lot more practical to store it in a safer place. Also in terms of character, robots will not resemble humans, but instead be specially adapted to carry out difficult or dangerous tasks.

What computer specialists have not yet been able to realize, is how to create intelligence similar to that of a human being. The brain of a human is the unique result of millions of years of evolution and it is particularly apt at recognizing patterns and identifying links. Most investigators believe that pattern recognition requires a huge memory reservoir and an extremely high processing speed. However, one might suspect that the problem is merely that we don't yet know how to write such programs so that they operate efficiently. Whichever is the case, progress is slow. But artificial intelligence, even intelligence that is superior to that of man, shall become reality, of that I am certain. And it probably won't take that long, either.

The path to true artificial intelligence could be laid out as follows, though various scenarios can be imagined: first, systems based on a certain expertise will be built, which are in fact gigantic databases. Doctors will build expert systems in which all medical conditions, symptoms, diagnoses and recommended treatments known to man can be stockpiled. The problem is that the method used requires an expert to spend a few years creating a program for one narrow area of medicine, but a number of successful experimental programs of this type have already been created and tested in the past 30 years (an interesting example is 'digitalis', a program designed for assistance in pharmacology and health care).

And then lawyers will want to examine articles of law and previous verdicts quickly and efficiently, and also in relation to technology there will be all sorts of expert systems that will provide the required solutions. Motorists will have access to GPS systems containing maps from all over the world. Expert systems of a higher category exist already: mathematicians have developed programs that can solve complicated equations, using formulas for integrals and other theorems known from literature. Sooner or later someone will have the brilliant idea of combining and linking all these expert systems, together with encyclopedias and databases — this is already happening on a smaller scale. Add to this faster and increasingly efficient search engines that recognize links and independently formulate answers to questions posed, recognize the human voice, and are able to answer the questioner in his or her own language.

What has started as a decent aid for the elderly will grow into an indispensable tool. Sooner or later we shall discover that these expert systems, now located everywhere in the world, appear to contain intelligence. In a sense, the Internet will become intelligent itself! This is a subject of much speculation. Not one but millions of computers are connected to the Internet, forming a gigantically large memory reservoir. Already, search engines like Google can immediately and elaborately answer questions such as *"what is the geographical distribution of the swamp snail?"*. Soon you will also be able to pose questions such as *"my knee has been hurting for the last three days, why could that be?"* — actually, you can do that already, but if you try that now you will be connected online to a real doctor of flesh and blood. That is not the intelligence I am talking about. I think that at some point in the future the Internet will take over certain functions now fulfilled by experts, such as those of doctors. The computer screen will then talk to patients and will far outperform the human doctor, including the online one, because of its perfect knowledge of medicine.

An intelligent Internet is not enough. Sooner or later we will require individual computers that are intelligent, and I believe this will become reality. I know fellow physicists who are of the opinion that for fundamental reasons a computer with 'consciousness' cannot exist. However, judging from the arguments they make, I must conclude that their arguments are based on emotions and not on facts. The computer I am working with at present most certainly has something like consciousness — admittedly below par — acting as it often does solely of its own accord and doing things I don't understand, as if it has a will of its own. No, real intelligent computers are certainly coming, even though I can't even make an approximate estimate as to how long this will take.

If you are asking yourself how many different facts a person needs to consider in order to answer a randomly difficult question, or to discern a certain pattern, than it appears as if the sheer amount of memory elements of our current computers is not that far removed. However, there is currently a complete lack of software that enables smart computer programs to compare data in such an intelligent manner. Nature itself has spontaneously developed a solution for this problem — the human being — but it has taken millions of years to accomplish. We have only been trying to mimic this accomplishment for a couple of decades.

Would it really be possible in the future for computers to reach calculated and premeditated decisions, in which social factors, human emotions, morality and intuition are taken into account? And how about humor? Conscientiousness? Irony? These elements are usually presented as typically human traits and often prognostics state that computers will never understand any of them. Well, I beg to differ. Our human emotions practically always have a biological background and are very easily explained and understood.

I am convinced that computer programs of the future will not encounter any problems understanding them either. Computers will not actually experience these feelings themselves, because they do not reproduce biologically, but they will understand them and programmers will succeed in creating software to enable computers to understand how to handle us humans, strange creatures as we are. I find it difficult to foresee all the possible outcomes of such developments.

chapter 6

Robots

ANYONE WHO THINKS about artificial intelligence, envisions robots. According to science fiction, these are machines that resemble human beings, by walking and talking just like us. But what we really need are machines that are *not* like us; ones that can operate in environments that are dangerous or uncomfortable for humans. They should be controlled remotely, so they won't have to carry their own intelligence inside those little round containers that house their optical detectors.

Robots have been in existence for a long time, in various forms. They don't have any real intelligence, because this is way beyond our abilities to produce. So they are controlled remotely or are pre-programmed to execute specific tasks. The most wondrous examples are the unmanned spacecraft that are currently exploring the far corners of our universe. Motorized vehicles are driving around on Mars, directed from Earth. Where do the boundaries lie?

A lot of developments are still possible in this field. As computers diminish in size, robots will too, and that makes their application for various simple tasks more diverse. Think of the vacuum cleaner robot. A vacuum cleaner robot that I saw recently royally botched its task. It didn't even come close to cleaning those hard-to-get-to corners, and it didn't really understand the layout of the floor it was supposed to cover. To accomplish its task properly, it needs a wireless connection to a larger computer in the home and it should be decked out with various add-ons, to reach every nook and cranny. It should empty its depository itself and reload its battery when necessary. Even better, it should only have to be told what to do and how to do it once, and if it accomplishes its task soundlessly as well, then the residents of the house would surely begin to appreciate having such a robot.

But more important applications can be envisioned. Presently, a road has to be dug up to lay or repair television-, water-, or gas cables or to reach sewage pipes. It would be a lot cheaper to use small robots to dig, as they would only need to resurface at the beginning and at the end. This development could make a real difference to community finances because the need for underground pipes, such as to accommodate fiber-glass cables for communication or computerized networks for the transport of goods, is only going to increase over time, particularly if they can be made cheaply.

If digging robots are able to carry out their work well and economically, then we can consider larger tunnels, such as for metro lines. In my country, the Netherlands, the problem remains that we are dealing with soft clay or sand, which do not provide any solidity; tunnels have always been dug from above, otherwise they cave in. I wondered whether it would be possible to have a small robot dig a structure resembling a skeleton, which we fill up with cement. This would create a strong capsule for a large tunnel, which could be hollowed out without the danger of collapse. But I have walked away from this idea, because I foresee unsurmountable problems with access and waste disposal. In the meantime, digging robots are becoming smarter and smarter, and the people above ground will no longer be bothered by the construction of tunnels. I foresee an increase in underground infrastructure: highways, parking lots, goods transportation — a lot of that could be done underground. Indeed, the city of Amsterdam has been contemplating the idea of creating a second city, entirely underground, consisting of highways, parking lots, shops and more, all beneath the old town itself.

The future also looks promising for unmanned airplanes or 'drones'. Such flying robots already exist, mostly equipped with cameras. These robots are frequently experimented with and, for now, mostly used by the military, but their action range is currently limited. As it is, an important law of physics says that smaller organisms fly much more easily than larger ones. This can be seen clearly in living organisms: small animals have a lot less trouble getting off the ground than larger ones. Therefore, once miniaturization sets in, we can expect to see lots of small flying robots.

In one of the science fiction stories I never wrote, these flying robots become popular children's toys. Equipped with cameras, they can travel anywhere and they become a hindrance and a threat to personal privacy. This results in a greater demand for all sorts of detection machines that keep these annoying peeping toms out of shower cabins and changing rooms — not a problem, because these little detectors will be widely available for a pittance.

With the emergence of nano-electronics and the possibilities of various devices of submicroscopic dimensions, robots can become very small. This is what Feynman was talking about in his famous lecture as well. These tiny robots will become particularly important in the medical sciences. Surgeons will no longer have to open up the patient's body to carry out an operation, but inject one or multiple robots into the bloodstream. Arteries and intestines could be surveyed and treated from within, and tumors would be detected much earlier on. These tiny robots could be driven by strong magnets from the outside. But there is one complication that is often overlooked. As I explained in Chapter 3, larger eyes can see more than smaller ones. If it becomes possible to manufacture minuscule robots, they will either be blind or would see only marginally. They would have to rely on touch, or perhaps a remote computer could combine the diffuse signals of a number of robots to create a usable and visible depiction of the inside of the body. Directing robots would be a complicated task, but far from an inhibitive one, as advanced, specialized computers will be readily available.

In another chapter of the science fiction novel I never wrote, large numbers of robots are injected into human beings to enable early detection and treatment of any conceivable affliction. Other fantastic possibilities are created if nano-technology is combined with another important branch of biology: genetics. I will devote myself to this subject in Chapter 14.

For the time being we are confined to robots that are controlled remotely, as opposed to robots with an independent brain. This will probably change in the somewhat more distant future. It is entirely possible that, where the distances become too great, such as with space travel, there will be a need for robots that do not require remote control. This topic will be readdressed in Chapters 12 and 13.

chapter 7

Victoriamaris

With a degree in naval engineering, my father was a senior executive for a shipbuilder on the Wilton-Feyenoord wharf at Rotterdam harbor. The company was tasked with the construction of enormous cruise liners that were becoming very popular at the time and which enabled people to sail from the Hook of Holland to New York in ultimate luxury. They were floating palaces. The flagship of the fleet was christened New Amsterdam. After that, cruise liners were given similar sounding names of Dutch towns, such as the Volendam and the Maasdam, and subsequent ships were given names of non-existent cities. My father worked on the construction of the Maasdam and the Rijndam, which were launched in 1951 — the Rijn (Rhine) is an existing river, but there is no town called Rijndam. After that came the Statendam and others.

The floating palaces taunted the imagination. Where could this lead in the future? Could whole cities be built floating on water? Could whole communities remain permanently at sea?

In my fantasies, I imagined the rise of world cities built on floating pontoons, trophies at sea. In yet another of my never-written science fiction novels, one of those cities was named Victoriamaris. It floated in the middle of the ocean, where

The Rijndam, 1951.

windmills and waves competed to generate most of the required energy. The climate was easy to regulate; if it became too cold or too hot, the entire float was maneuvered into a more northerly or southerly position and as this did not require great speeds, the thrust of the large sails would suffice.

The vision of such super-houseboats was also conceptualized by Professor Frits Schoute. He discussed his ideas at his departing lecture when his position at the faculty of Electrical Engineering, Mathematics and Computer Science of Delft University came to an end. Environmental friendliness was high on his list of priorities and his first 'ecoboat' is a large floating pontoon with a laboratory and windmills. In addition to the wind, energy is generated from the waves, using cylinders containing water moving back and forth. This energy is needed to drive the propellors, to create potable water and to make water suitable for other domestic uses. Slowly but surely, larger pontoons are created and added to the existing structure. A wall of large buoys protects the city against turbulent waves and with the tempering of the ocean a lot of usable energy is created. The buoys are attached mechanically — or perhaps magnetically — to the city. It is not a problem if these buoys run astray in heavy weather; we retrace the buoys after the storm and repair the connections.

In Professor Schoute's vision, houses, schools and shops appear — in other words, a complete village. Seaweed is grown in suspended ponds — not for direct consumption, I hope — and the houses are heated with a heat pump. The village is mostly inhabited by home workers. A hydrofoil is used to travel between the city and the mainland.

Technologically, all this is possible, but whether we actually want to live in the confined environment of an artificially-created island — and particularly, whether we want to pay for it — remains the question. And whether it is safe is also a moot point. Tsunamis pose little problem, because if the water you are in is deep enough, the wave would pass right underneath the floating city mostly unnoticed, but it can be quite stormy at sea and such a floating monster is too slow to maneuver away from a pending storm. An iceberg, such as the one that sunk the Titanic, is also not much of a worry. With the aid of information technology we would be able to locate floating beer cans, let alone an iceberg.

But anyone dabbling in science fiction who takes himself seriously must come up with more than a few floating houses at sea. Would a city floating in the air be possible too? It sounds bizarre, but the laws of science allow for a lot of leeway in this regard. Decent-sized hot air balloons of, let's say, a few dozen meters can carry several people and a large platform is easily floated as long as it contains a sufficiently large mass of gases that are lighter than the surrounding air. We may choose hot air for this, but also different types of gases such as helium or hydrogen. Heated air, it seems to me, is the easiest option. We could build pontoons on, underneath or between enormous zeppelins. The bigger the creation is, the easier and cheaper it will be to thermally isolate the hot air from the outside. After all, larger constructions can preserve their heat more effectively, because the surface area which is exposed to the outside is smaller in relation to its volume. Stability could be an issue, but here also, the bigger the floating structure, the easier it will be to stabilize. Other problems might be more difficult to address: in stormy weather, wind sheer and turbulence may become a threat to large, light constructions.

Barring such inconveniences however, details can be filled in as desired. For example, in mountainous areas, floating terraces could be built for farming or livestock breeding. Free-floating structures could be many kilometers in size and house permanent communities; there would be grasslands for fresh produce and cattle, and shops made out of feather-light material. Engines would be desirable but not strictly necessary, because it is possible to steer these structures in various directions by adjusting their height, such as is done with hot air balloons. The total expenditure, however, would be astronomical, and the motivation to build a floating city would be minimal, so I don't see any real future for such 'pie in the sky'. But it *is* possible.

chapter 8

A Malleable Earth

CITIES FLOATING ON water or in the air, and all of them as ecologically friendly as possible. Will this be a realistic future, inspired by science fiction? To be frank, I doubt whether those beautiful ecological ideals will be attainable, as we can't even stop destroying the environment we're in, let alone change our ways in a habitat that we have not yet been able to build. Rather, the mere existence of floating cities in the future would prove the contrary, and tell a gruesome story of the continued waste of huge amounts of energy from burning fossil fuels; our climate irrevocably changed, the sea level risen all over the globe, and my beautiful country, the Netherlands, under water. Because we were not able to change our wasteful ways, building floating cities was our only alternative to continue living on Earth.

Will this actually happen? Should we just give in to the salty waves and accept that our nations will be swallowed up by the sea? Is it possible to persuade the world population to permanently change their habits of consumption, waste less and pay more for energy? President George W. Bush of the United States did not appear to warm to the idea. The American people must be allowed to stick to old traditions and waste as much energy as they like. Of course, New York won't be usurped by the sea; surely technicians will find a solution to this problem before that happens, or so it is presumed. And if the solution is going to cost money, it won't be paid for by the American government, if the present stance of their negotiators at today's climate conferences is anything to go by. Angry gestures towards the Americans and the Chinese are unlikely to generate a more positive response. But they are not the only inhabitants of this planet who are spoilt and unwilling to pay taxes and contribute to the greater good. The Dutch will have a go at it themselves. Fortunately, the Netherlands does possess a lot of in-house knowledge about water management and we *do* pay taxes, and more than the Americans.

It must be said, however, that climate change is nothing new; the climate of this planet has always been fluctuating, due to natural causes. The ice ages, the Pleistocene period, lasted almost two million years and ended approximately ten thousand years ago. After that, during the Holocene period, the ice receded and the sea level rose to its current level. To date, it has risen slowly, about a centimeter every hundred years, but during the last century a quicker rise has been detected. Paul Crutzen, who won the Nobel Prize in chemistry for his study of chemical processes that occur in the Earth's atmosphere, suspects that we have entered a new era, which he calls the Antropocene period. It is the period during which humanity causes great changes in the earth, water and air. Industrialization developed quickly and was accompanied by the burning of fossil fuels on a massive scale. It is perfectly conceivable that, in response, the sea level will now rise at an unprecedented speed. In total, this rise will not be as gigantic as the events of the distant past, but those events took many thousands or perhaps millions of years to take place. Now, changes might happen in only a few centuries. If so, this will be a huge problem for the low-lying Netherlands.

Through the centuries, the Dutch have learnt to build dykes and flood locks. The New Waterway Storm Surge Barrier located in the Oosterschelde in the Netherlands is a perfect example. It is only a matter of time before we are going to have to far outdo our previous accomplishments if we want to protect our country against floods. The old, trusted methods that we use now won't hack it. Take the building of dykes. The Dutch ground is soft. If you want to raise a dyke by one meter, it has to expand by ten meters in width and what is worse, it will sink 90 centimeters into the earth in only a matter of time. And don't forget the groundwater levels. We would have to use increasingly bigger pumps to avoid salinity of the ground water. If the sea levels rise, so will the rivers, so the river dykes will need to be raised as well.

The financial interests to safeguard our country are enormous, so in all likelihood bigger dykes will be built. But how do we avoid the degeneration of these dykes? Could a dyke be constructed out of very light material? No, that would be too dangerous because it might be pushed up by the water, resulting in water flowing through and underneath the barrier. So, I am thinking of a novel kind of construction: large blocks of concrete made hollow on the inside. On purpose they will not be made watertight, so that at the waterside, the water can flow in and out. This way, the dyke controls its own weight: heavy when the water levels are high and light when low, it won't sink or slide over time.

Water engineers, much better versed with such issues than myself, will of course busy themselves with this problem, and may well come up with much better solutions than mine. I don't think we will sit back and sacrifice our country to the sea. A

Diameter of a dyke, built with hollow concrete blocks. At high tide the concrete blocks fill up with water, which makes them heavier. This improves their ability to resist the water pressure. At low tide the water drains from the hollow concrete blocks, minimizing the risk of the dyke sinking into the ground.

technological masterpiece will be devised that will protect us. The enormous expenditure of it will have to be recovered by selling our expertise abroad. After all, every coastal area of the globe will face exactly the same problem.

Would it be possible to halt climate change, which is the cause of this and other troubles, taking recourse to technological wonders of a different type? The prognosis for the short term, by which I mean the first few hundred years, are not good. What was only ten years ago a mere potentiality, is now, excuse the pun, water under the bridge: our climate is changing. Those changes are indeed happening fast and in all likelihood they are the consequences of human actions. The sea levels will continue to rise, and fast too. Estimates vary wildly — climatic predictions are notoriously difficult — so that all we can say is that climatologists suspect the rise of the sea levels to be anything between eight and 90 centimeters between 1990 and 2100.

That may not sound like much, but in due course this could grow to be disastrous. Climatological environments are not stable. Here is why: various large land masses on Earth are covered by gigantic amounts of ice. The largest ice mass is that of the Antarctic, which is covered by an ice sheet up to 4.5 kilometers thick. Another large ice mass is Greenland, but its surface is much smaller than that of the Antarctic. If the ice in these areas melts, its meltwater will flow into the sea, and that is such a vast amount that the sea level will rise quite significantly.

Ice is not only located on land masses, but also in the polar seas, notably in the Arctic. That ice will also melt, or at least partially, but will not cause any significant rise in sea levels because it floats on water[c]. Look at a glass of water. If ice cubes in a glass of water melt, the water will not overflow, but if ice cubes on a tray melt, and the water flows into a glass, than the glass could well overflow. So, ice melting off land masses — that's what we need to watch out for.

---

[c]Also, ice consists of unsalted water.

Ice melts when it comes into contact with warm air, but not when it is exposed to sunlight. That is because ice is white and reflects most of the rays. However, as soon as rocks are exposed, the surface absorbs a lot more heat. Then the temperature rises and the melting process accelerates. Therefore, as soon as land becomes exposed in the Antarctic or Greenland, then we are in trouble. The melting process can no longer be halted. In winter those places are covered by snow, but this melts in summer. Eventually, a new equilibrium will be reached. Maybe the ice will not disappear completely, because of the increased humidity of the air, which will cause an increase in precipitation, and hence, more snow. Where this balance will lie is too difficult to tell just yet.

What we do know, is that there have been several periods in geological history where there was no ice at all on the South Pole. Many millions of years ago, the continents were positioned differently than they are now. Instead of the Antarctic, Australia was at the South Pole. But there was no ice. Dinosaur skulls have been unearthed that have huge eye sockets. At the South Pole it can be dark for months on end, so these animals would have needed to be able to see in twilight or even with not much more than the little bit of light shining from the stars.

During this period, seawater levels were much higher than they are now, and that could happen again. It has been calculated that if all the ice on the Antarctic melts, the water levels would rise more than 60 meters. The Netherlands and New York would cease to exist. This cannot happen quickly — it would take thousands of years.

The process has already begun, slowly but surely, but no one knows if we can turn back the tide, so to speak. The ice on Greenland is already melting at an alarming rate. It could add eight meters to the sea level. The cause for the rise in temperatures on Earth are greenhouse gases, particularly carbon dioxide, but there are other gases that contribute. Most of these have been emitted unintentionally into the atmosphere by human activity, such as methane, which stems from the excrement of our humongous livestock herds. The amount of methane in our atmosphere has doubled in the last century. Carbon dioxide (CO_2) is, however, the main cause of the recent climate change, and we continue to produce more and more of it. It is formed whenever fossil fuels (such as oil, coal and gas) are consumed, which are the main ingredients of all energy sources upon which our present civilization currently depends.

Methane and carbon dioxide are so-called greenhouse gases, named that way because they act just like the glass in a greenhouse. They are transparent for visible light, which has very short wavelengths. The Sun pours most of its energy upon the Earth in the form of visible light, and when it reaches the soil it is converted into heat. The Earth in turn radiates this heat back into space, but, because it is much cooler than the Sun, its heat is emitted into space as infrared (invisible) light. Infrared light, which has much longer wave lengths than visible light, is partly trapped between the land and the carbon dioxide clouds. In doing so, small amounts of such gases can effectively heat up our planet.

Ironically, we have been able to lessen the emission of sulfur compounds, another pollutant with which humanity has contaminated its environment. Achieving this reduction was important. Sulfur is bad for our health and causes acid rain. Sulfur dioxide, however, also appears to *cool* our atmosphere, because it creates something like a cloudy veil consisting of acidic drops. These droplets reflect visible sunlight, which is why they act as coolants. However, we don't want sulfur in our air, leaving us without the option of using this convenient side effect, bringing us back to the problem of reducing the amounts of carbon dioxide, the biggest cause of trouble.

All sorts of suggestions have been made to reduce the emission of carbon dioxide. However, there are so many people driving cars, and so many power stations running on coal or gas, that we are talking about huge amounts, and the current proposals are not likely to make a significant dent in the level of emissions. Nature has her own methods: the coastal areas will be submerged, the world's population will diminish because of large scale loss of habitable land, and this will automatically cause a reduction in damaging emissions.

Geologists know what that could mean: it is possible that a period occurs when there is very little carbon dioxide in the air. This would cause the Earth to cool off and the ice layers to get thicker. Ice could even cover the entire planet. There is evidence to suggest that between 750 and 580 million years ago, there were several — albeit short — periods of time where the entire Earth was covered by thick layers of ice, including the areas at the equator. Our planet was one big snowball. Unlike what one might think, such a snowball could not last forever. The activity of plants, or whatever vegetation existed in these bygone eras, would have ceased, so that conversion of carbon dioxide into oxygen came to a halt. At the same time, volcanic activity slowly increased the amounts of carbon dioxide and methane back to the levels they had before, resulting in a gradual retreat of the ice within a couple of million years.

In one of my never-written science fiction novels the Earth threatens to become a snowball and the world population must once again burn large volumes of oil, solely to increase the level of carbon dioxide in the air. Anyway, we are nowhere near this situation at present. The moral of this story is clear: the climate is not stable and we have to intervene. It concerns enormous amounts of carbon dioxide and whichever solution we decide upon, it will cost lots of money. The most expensive solution, however, is to do nothing, because this will lead to the flooding of all coastal areas — probably only after all sorts of money-devouring projects have been carried out to reinforce our dykes.

So, first, all sorts of catastrophes will have to occur and only then will political powers start to pay attention — including the American president. But what to do? I don't profess to know everything. What will happen exactly and what the best solutions will be, I don't know. I can only argue passionately to approach the problem scientifically. ALL proposed — and somewhat realistic — solutions will have to be considered, post haste.

First, we have the option to save energy. This does not have to happen at the expense of life's little luxuries, but could be realized simply by making our appliances more economical. I find it incomprehensible that most appliances continue to have standby lights, which absorb dozens of watts. That should be milliwatts instead! Computers that are left running while not being used for hours on end should run on very little electricity. Lights should turn themselves off if no one is around. The need for transport can be diminished by gradually improving electronic communication, and so forth. My belief is that modern information and communication technologies can be enormously efficient in reducing our energy consumption by large factors, without any loss of comfort.

Further, the Sun and the wind are widely known to be alternative sources of energy. Let me start with wind energy. It is unlikely that there is enough energy in the atmosphere to satisfy all our needs solely through wind farms, but cyclones intrigue me. There is evidently a lot of energy in the lower levels of the air, which become warm and humid because of superfluous sunshine. As soon as a certain level is reached, the layers of the air become unstable, and the lower layers want to trade places with the higher, much cooler and dryer layers of air. That is accompanied by much bluster, resulting in what we call a hurricane.

It would be very difficult, but could we harvest that energy before it reaches that fateful level? I am thinking of enormous chimneys, several kilometers high and hundreds of meters wide, for instance to be built in dry, desert-like environments. Warm air would be sucked in from below and rise through the chimney to the top, cooling down in the process. Large wind turbines harvest the energy from the raging winds and in addition, these vertical movements generate rain. In the Caribbean seas, the air is not only warm, but also humid and therefore even more energy rich (moist air is *lighter* than dry air). The chimneys in these regions would probably not have to be as high to generate lots of energy.

Such a chimney would be the artificial eye of the cyclone, but this cyclone can't move anywhere. We erect hundreds of such chimneys and by controlling the supply of air, we can turn them on and off, and in this manner, hopefully, influence the behavior of a real cyclone. Eventually, there would be so many chimneys that there would be too little energy left for real cyclones. We would then have killed two birds with one stone: an abundance of energy and no more dangerous hurricanes. Naturally, there would be lots of opposition to this system: this energy source strongly depends on the seasons and the legal ramifications of the first failed attempts to stop a cyclone would be unforeseeable: any hurricanes that did occur would be blamed on our project! Yet this is just one of these intriguing high-tech ideas that I love to speculate about.

Solar energy, however, is plentiful. Usually this means solar cells, which convert direct sunlight into an electric current. There is but one big problem with this: solar cells are not economical. A solar cell is expensive and vulnerable. The latter is crucial. Everyone knows that exposure to the Sun's rays causes damage. Solar cells are refined electronic circuits that convert the harmful photons into electrical currents. They don't last long, as real life often shows.

I would actually favor the use of large mirrors, which concentrate sunlight onto generators, and convert the Sun's heat into energy in a more conventional way. One of nature's laws states that a source of light that has a given temperature can heat up another object close to the same temperature, but not higher than its own temperature. Now the temperature of the surface of the Sun is somewhere near 3000 degrees, so it is possible, in principle, to generate similar temperatures in the focal point of a mirror, but never higher than 3000 degrees. In practice no more than about 500 degrees can be reached, but this is more than sufficient for very efficient energy transformation. The problem is that huge areas are needed to generate enough energy.

An objection sometimes voiced against solar and wind energy is that there isn't enough of it when you need it and too much when you don't. It should be clear that good methods need to be invented to store this energy. There are so many competing solutions, that as a scientist it is difficult to predict which one will prove to be the most economical. Hydrological storage might be an option, where water is pumped from low to high, or perhaps by using flywheels. Perhaps efficient electronic storage methods will be invented, but storage in the form of heat is also possible, by using efficient heat pumps.

An important element of the discussion is the role of nuclear energy. Nuclear energy does not produce greenhouse gases. Organizations such as Greenpeace continue to oppose nuclear energy because of ethical arguments, but their standpoint is disputable, at least from an environmental point of view, particularly if by 'environment' we mean flora and fauna. Even if an occasional nuclear disaster is to occur and large amounts of radioactivity leaks out, the flora and fauna suffer very little. A few plants and animals would of course die, but that would only be a small percentage, and they would quickly be restored.

What people don't want to accept is that several hundreds of us might die of cancer. Sometimes thousands or tens of thousands of victims are quoted, but remember that radioactivity also occurs naturally and many of those cases can actually be attributed to natural causes, and not to human action.

We would of course want nuclear energy without any accidents, producing the minimum nuclear waste and, if at all possible, without byproducts that can be used to manufacture those horrible weapons. My argument would become a bit too technical if I were to go into too much detail, but suffice to say that there are various solutions to producing nuclear energy, even though these will inevitably be opposed by anti-nuclear campaigners with their almost religious fanaticism.

A fervent supporter of a new generation of nuclear reactors is particle physicist Carlo Rubbia, Nobel Prize winner of Physics in 1984. His proposition concerns reactors that work with a particle accelerator, using a design that should be much safer than conventional models. The procedure is based on nuclear fission, though not of uranium or plutonium, but of other heavy chemical elements such as thorium. The use of such raw material creates waste that is less radioactive after a few hundred years than the original. The reactor stops working when the particle accelerator is switched off and can therefore not explode. It is not possible to create weapons with thorium and technically such a reactor is feasible, even in the not too distant future.

Aside from the above, nuclear *fusion* is also an alternative. This is a very different source of nuclear energy than what is currently being used, which is nuclear *fission*. While heavy atomic nuclei are being split into smaller parts with the conventional method, it is very small species of atomic nuclei that are fused into larger ones in a fusion reactor. Also with this process, little or no radioactive material is produced and the fuel — essentially just water — is available in abundance. A fusion reactor is, however, extraordinarily difficult to realize because reactive gas with a temperature of millions of degrees Celsius held in an extremely complicated magnetic arrangement must somehow be controlled.

In the near future, it will hopefully be possible to demonstrate that such a reactor might nonetheless be used as a generous energy source with the construction of ITER, the first fusion reactor that will provide real electricity. A large laboratory is located at Cadarache, a town in the south of France. This is also where the construction of the reactor will take place. It is a location that emerged as a compromise after lengthy disputes between the participating nations.

One of my colleagues remained conspicuously skeptic. "*It may be*", he said, "*that they will succeed in controlling hot gases in strong magnetic fields, but there will be a lot of radiation. That radiation will disastrously affect the coating of the walls that must be there to contain a high vacuum. It will have to be repaired or adjusted very frequently. How are they going to realize that?*"

Well, considering the economic interests at stake here, I think financially viable solutions for this technical problem will be found. In any case, it is extremely important to begin to invest in these potential alternative energy sources, even if all the kinks have not yet been ironed out. Just like the science fiction writers, I have enormous faith in human ingenuity.

Unfortunately, no economically viable application of fusion energy is expected to be realized before 2060 and until then we will have to find other alternatives. When in the 1960s work on fusion energy was started, it was predicted that it might take another thirty years before nuclear fusion would be viable. The fact that in 50 years that term has been prolonged to another 50 years, arguably does not look promising for the future. But in this matter too, the advance of science and technology will make more things possible.

Will the Sun, wind, nuclear fission or fusion provide enough alternative sources of energy in time? Most likely, the conversion will take too long and will be a mere drop in the ocean. Even more forceful measures will be necessary.

Let me suggest another solution, even if we are far too late with it. There are other aspects of the world's climate that we would like to change. I have made numerous trips through deserts and I always wondered: if people here had access to plentiful supplies of fresh water, would there be dense forests? Lavish and fertile soil for agriculture and livestock? Would there be more space for humans to live? Thus, we ask: would science and technology not be able to create large quantities of fresh water? Would not the benefits far outweigh the expenditure? What would be the consequences for the world's climate if there were a huge increase in the fresh water reserves in most of the world's deserts? Would this not also cause an increase in precipitation? And what would happen, if a rich flora and fauna emerged, that would convert large amounts of carbon dioxide from the atmosphere into vegetation and deposits? Would we not kill two birds with one stone? We would create a huge area suitable for habitation, enable lavish consumption and in addition extract carbon dioxide from the air. Of course, we wouldn't want to convert *all* deserts into farmland and forests — we should always preserve indigenous desert, if only to preserve the biodiversity of our planet — but surely we can do with a little less?

Water has always intrigued me. Israel as well as Jordan sit by and watch the Dead Sea dry up. Colleagues of mine in Israel came up with the idea, some time ago, to dig a canal that connects the Mediterranean Sea to the Dead Sea. The Dead Sea is situated at the lowest point on Earth. The water level is four hundred meters below sea level. If the canal were to be dug, water would flow to the Dead Sea with enormous force, and a lot of energy could be generated using turbines. In addition, water levels would be restored to the original levels and condensation would decrease the area's drought. This plan was swept off the books for political reasons, but a similar plan has surfaced more recently: Israel and Jordan would together dig a canal from the Red Sea to the Dead Sea. The distances are greater, but the result would be the same. Whether the plan will go ahead, however, depends on factors apart from science.

In the meantime, the Arab world also suggested the most amazing ideas. There have been plans to tow icebergs from the Antarctic to the Red Sea. That would generate huge amounts of fresh water. True, the journey would be so time-consuming that most of the ice would melt along the way, but there would still be more than enough left.

Would such a scheme really be economically viable? I contend that we should build enormous desalination machines. Ideally, they would run on solar energy as opposed to oil and gas. Sea water is heated; it evaporates and condenses into fresh water. We could also pump water through filters using machinery driven by solar energy. In the longer term, I see another possibility: genetically-modified vegetation filtering water with the aid of sunlight.

While I write this, Spain is coping with serious drought problems. This country should take the lead in developing large-scale desalination facilities. They are expensive, but beneficial to the atmosphere in the long run. Other countries, and in particular the European Parliament, should provide support to Spain. The process of transporting water from the sea on to the land would reduce the sea level only slightly, but the beneficial effect from the increase in absorption of greenhouse gases would reduce the sea level to a much greater extent. Who knows.

I imagine other large-scale projects that could influence the Earth's climate. One of the reasons for the minute amount of precipitation in parts of Africa is because the Atlantic Ocean is so cold and therefore only sparsely evaporates. Couldn't we construct shallow basins where the Sun can easily heat up the salty water and cause it to evaporate? The only way to find out if such an idea is feasible is to subject it to powerful computerized calculations. With these calculations, we would know where to pump the salty water, so as to enable the Sun and the wind to do their work.

There are definitely ways to convert sparsely vegetated areas into densely covered forests. In 1994, an entrepreneurial economist Gunter Pauli founded Zero Emission Research and Initiatives (ZERI), an idealistic organization that aims to increase the productivity of underprivileged regions significantly through simple technologies, without infringing upon the environment. A beautiful success story concerns the savanna in the eastern part of Colombia, where nothing would grow. If you could grow a forest there, it was argued, you should be able to grow one almost anywhere on this planet.

The biochemists of this organization found out why no conifers would grow in this area. The soil had a pH value of four, which indicated that the soil was too acidic and furthermore the sunlight was too bright, which implied that the seeds would not have any chance of survival. The problem was solved by first planting bushes that could resist the Sun reasonably well, paired with a fungus that would decrease the acidity of the soil.

In the shadow of this pioneer vegetation, conifers were planted. Presently, the trees grew without any problems, and 8000 acres of savanna turned into a forest. As a result of this forestation, precipitation increased as well, which enabled palm trees to be planted. The economic value of the region increased dramatically. Affecting climate change and contributing to the fixation of carbon dioxide; it doesn't get any better. This is a great example of science applied well.

It is important that we do not burn the organic material we produce with these sorts of projects, or allow it to decay, but

The artificially grown tropical forest in the savanna of eastern Colombia.

that we find a lasting purpose for it, in order to avoid the carbon dioxide contained in the material from returning to the atmosphere. If we produce high quality wood, we could use it to build homes or furniture. If we run electricity generators on biologically-produced fuel, we must not make the mistake of assuming that any carbon dioxide emitted in this process is harmless. We should try to store this carbon dioxide underground, just as engineers plan to do with conventional plants running on fossil fuels.

Paul Crutzen recently launched an astute and clear-headed proposal. He suggests a very different approach to global warming and the rising sea water levels. I explained earlier that sulfur in the air is hazardous to our health and that of our plants, but that sulfur compounds actually lower the temperature of the Earth. That is because high up in the atmosphere the Sun heats up the sulfur, which in turn causes water droplets to appear. These droplets reflect sunlight particularly well in the visible part of the spectrum.

The cooling effects of such droplets is well-known; they are also seen in the contrails left by high flying airliners. These high altitude artificial clouds have a definite cooling effect, as noted during the few days after 9/11 when all airliner traffic over the USA was suspended and there was, consequently, a large measured increase in the total insolation over the whole country!

It transpires that the effects of the sunlight reflecting off sulfur droplets are much higher if these drops appear high in the stratosphere, simply because sulfur remains afloat much longer here: two years, as opposed to mere weeks at lower heights. The annual emission of sulfur is decreasing from what it was in the past, 160 million tons, and this is fortunate because we definitely do not want that poison in our atmosphere. But if we were able to spread a mere two million tons of sulfur high up in our stratosphere, that would cause enough cooling to neutralize the current effects of our carbon dioxide emissions. The contribution to acid rain would be negligible and our health will also remain unaffected.

How would we get the sulfur into our stratosphere? With a large cannon, was the first thought. Large cannon balls of sulfur would be fired off into the air and explode once the stratosphere is reached. "Criminal!" I already hear people yell, but this is a sensible proposal that deserves further study because there may be even smarter methods that would put sulfuric gases to good use at great heights without us being affected by them at lower heights. Alternatively, it could be decided that airliners must burn high-sulfur fuel in addition to normal fuel — mixed into the jet engine's fuel in flight — after the airliners have reached cruising altitude.

And you don't have to worry, because this research would be conducted by climatologists who would not undertake anything without thoroughly studying and tracking all the thinkable side-effects, good or bad, and who would not dare to go ahead with it without the proper green light from all political echelons.

We have to make a careful distinction here between climate and weather. The climate of an area is determined by a certain weather pattern. The weather refers to the daily situation. Today's science is crystal clear in telling us that only short-term weather predictions are possible. How short this term is varies depending on time and place. If the weather has been stable for a while, it is easier to look ahead, but sometimes the locations of high pressure areas and depressions are so capricious that a prognosis for even a few days is difficult. As our computer models advance, the reliability of our predictions will improve, but it would mean extending that term by only a few days.

Accurate, detailed predictions covering a longer period of time, let alone controlling the weather, is not yet in the cards. What do the laws of nature tell us about this? They imply that weather systems are chaotic. Every disruption, however small, can change the weather totally, given time. If even the smallest of butterflies decided to spread its wings slightly differently than anticipated, then the effects on our atmosphere would increase over time and after a few weeks or months our weather system would have changed completely.

Such changes are not predictable. But would an advanced civilization be able to exploit this? With large mirrors, its climatologists would manipulate the sunlight either to heat a place or to be blocked, to evaporate water or not, to heat up certain parts of the Earth or just the opposite. These super-scientists will have identified the place and the time that would have the largest effect on the equilibrium. The disruptions only need to be larger than the effects they can't predict, such as when a butterfly may spread its wings.

The required computers will be so huge, strong and quick that they might actually be able to foresee all the possible consequences of the disruptions they caused, so that they could direct the weather towards wherever it is desired, for example by redirecting a storm front towards an area where it could do little damage. Bringing about such a system on Earth, making it cost effective or even desirable is extremely unlikely in my opinion, but I promised you I would explore all the possibilities within the laws of nature. Contrary to manipulating a climate, influencing the weather is not likely to be practical even in the future; however, in principle, even that is indeed possible.

The conclusion of this chapter is this: we want to be able to influence the Earth's climate in the longer term. Today, we have air-conditioning and central heating only within our homes; in the distant future, we might be able to have better control over the climate of the entire globe, using our 'Big Science': a malleable Earth. We want to have a better grip on our climate to be able to apply it more effectively in line with our requirements. But it is obvious that we have a long way yet to go before no rain will fall on our parade.

Ultimately, our unstoppable population growth is the prime cause of climate change, and at the moment this is putting a greater pressure on our planet than it can handle. While this problem should certainly be addressed by sociologists, I realize that there is nothing harder than controlling human beings. Is it not true that an increase in wealth leads to a decrease in population growth? Perhaps then, science can indeed be employed to control our own population.

I am convinced that if left solely to science and technology, the standard of living can still be improved significantly, even if the world's population continues to grow; and even in that case, there will be scientific ways to control the climate. This will not happen soon, since scientists have hardly begun to scratch the surface of this gigantic problem, but if you want to speculate on the science of the future, this is a truly fascinating subject.

chapter 9

Flying Kites

THERE WAS QUITE a lot of wind. I didn't have to go to school that day. My kite was pulling hard at the string I was holding in my hand. The kite was pulling a bit too hard in fact, and every now and then it swerved dangerously. It wouldn't be the first time. Only a few days ago a gust of wind had pulled the spool right out of my hand and I had to run after my kite for three blocks before I got it back. Unfortunately, my spool had taken a different route. The wind blew my kite over houses, antennae and balconies, but it was now dangling above the busy main street of The Hague, where I grew up. If the string broke, there would certainly be an accident — motorists and bike riders would be rather surprised by the sudden appearance of a rope across the road.

My kite and I had drawn the attention of a policeman. He got off his bike and was walking towards me. "*Is that kite yours?*" he asked. Yes, it was. I couldn't deny it. After all, I was still holding the spool. "*Did you make it yourself?*" Yes, I made all my kites myself. Only my first kite was bought from a store, and that one wouldn't fly at all. But now I knew exactly what had been wrong with it. "*Isn't it great fun?*" the policeman asked. "*It reminds me of my younger days, when I often flew kites. I would send up little notes, along the rope towards the kite.*" Sure, I knew how to do that too.

The wind would blow them upwards. Only now there were a few too many knots in my string, because it had broken so often. The policeman got back on his bike and rode off. I reeled in my kite; it was a bit too dangerous here.

A lot can go wrong with kites. An article in a newspaper reported recently that an unfortunate kite flyer had lost control of a beautiful kite in the form of a dragon when its rope got tangled up in a tree. The kite hung half a mile up in the sky above Schiermonnikoog, one of the Dutch islands, and was posing a risk to local air traffic. Firemen were called to try to rope it in, but they couldn't reach the kite either. There was a military base not far away and, eventually, the kite was shot down.

Firemen? Shooting down kites? Why didn't they just use another kite? Anyone who has ever flown a kite knows that it is very easy to take down a kite with another one. Usually it is hard to avoid it! Kite fights are a common sport in India. They sell a special kind of kite string with little bits of glass embedded in the string. You try to get your string to cross against someone else's kite string then you move your arm up and down rapidly to try to cut through his string before he can do the same to you! The specialized kites are called 'fighting kites'. It is a well known sport. My kites' strings would also get tangled in trees, but I was always able to get them down.

I have given kite flying a lot of thought. How do you stabilize a kite? How should one be built to get it to fly as high as possible? I was astonished time and time again to find the numerous and various constructions that are possible, each of which aim for the sky and defy gravity. What I realized much later, was that while building the kite is important, the really vital part is the string. The stronger and lighter the string, the higher your kite will fly.

Wubbo Ockels, the first Dutch astronaut and one of my friends, had understood this much earlier on. Wubbo is also an enthusiastic kite flier. According to Wubbo, provided the string is strong enough, you could fly a kite up to dazzling heights. It would even be possible to fly kites high into the upper levels of the stratosphere. If we tie together two kites which are built so that one flies much higher than the other, they would fly in different parts of the sky according to the different wind speeds. These kites would then pull each other up and fly around until the wind velocities and direction are the same in both layers of air or until one of the kites brakes; probably the latter. Wubbo also believed that it was possible to pull a glider plane behind a motorized airplane, just like a kite, up way into the stratosphere. He actually tried to do this, but the winds were too turbulent and, furthermore, handling such long cables turned out to be no mean feat.

Would it be possible to use kites to generate energy from the wind? At the heights where kites usually fly, the wind speeds are often high and constant, much higher than just above ground where we build windmills. Wubbo thinks he can build power stations with kites. A large number of kites are attached to a long, closed circular cable, like flying wings. At one side of the cable the wind pulls the kites up, while on the other side that motion flips the wings so that, without having to pull very hard, these come down. The cable drives a dynamo and this way a lot of wind is transformed into energy. Thinking back to my concern when I was looking at my kite hanging above the busy main street of The Hague, I am afraid this will not be an easy feat, because miles-long cables are extremely difficult to control. I wish Wubbo the best of luck with his 'ladder mill'. And the cables? I will return to those later.

chapter 10

The Stars

WE HAVE LINGERED on the Earth for long enough now. We want to go up, straight into space! We have already put people on the Moon, and many unmanned space vehicles have found their way to other planets and moons in our solar system. A few of those are about to exit our solar system, but in astronomical terms, they won't get very far. Long before they get anywhere near the interesting worlds outside our solar system, they will cease to function.

What are our prospects? As long as humanity has existed, we have gazed up into space. First with the naked eye and then with all sorts of astronomical equipment. Ever larger optical telescopes gave us increasingly sharp pictures of the furthest removed corners of the galaxy, and even those are surpassed by the observations of ingenious radio telescopes. One of the first truly gigantic radio telescope arrays is the one in Westerbork, in the eastern part of the Netherlands, but it was followed by many others in many different parts of the world. Instruments that have been sent outside our terrestrial atmosphere tell us more about the infrared, ultraviolet, X- and gamma-rays that are emitted by celestial bodies, and about an intriguing background radiation of microwaves, whose origins are from regions of the universe much further away, from the time when the universe was only a few hundred thousand years old.

THE STARS

There are enormous objects in our galaxy, many times bigger than the Earth or the Sun, which emit unimaginably large volumes of energy. The most remarkable characteristic of our universe, however, are the enormous distances. Light, which moves pretty fast with a velocity of 300,000 km (187,500 miles) per second, takes hundreds, millions or even billions of years to travel the distances between the various celestial bodies we have observed. Our universe really is enormous.

That's why it may have been a bit of a disappointment for you when I tried to explain earlier that humanity will probably never get any further than the outermost regions of our own solar system, somewhere around Pluto. Our solar system — also called our planetary system — is that part of the galaxy where our Sun shines brighter than any other stars. It stretches from the planet Mercury to quite some distance beyond the planetary pair of Pluto and Charon. As a side note, both have recently been downgraded to the status of 'dwarf planet' and 'moon', respectively; 'to pluto' is now an official verb in the English dictionary, meaning 'to downgrade'. An example that is given: *"he was plutoed like an old pair of shoes"*.

The Sun also belongs to our solar system, but for obvious reasons we won't be able to go there. In astronomical terms, our solar system is about as big as our backyard.

As I wrote in Chapter 2, we are fortunate that our Earth is small and light enough for us to build rockets that enable us to leave the atmosphere. Once we are in orbit, a trip to the Moon or any other planet requires relatively little fuel.

The total acceleration that a space engine has to produce without external aid is an important concept that defines its potential. If the engine produces gases with an emission velocity of three kilometers per second, then every time the engine generates an extra speed of three kilometers per second to move the ship forward, the remaining mass of the entire system will have diminished by factor 2.72. This means that to create a speed of six kilometers per second for every kilogram, 2.72 × 2.72 − 1 (or 6.4) kilograms of fuel is required. And so forth. If one calculates the velocity required to reach other planets, then this proves to be highly dependent on the chosen trajectory and the desired duration of the trip.

One way would be to ride on the tide of gravity of surrounding planets and moons, so that spaceships are flung away by the larger planets, as it were. At this moment, there are at least five small spacecraft that are on their way out of our solar system. Until recently, Voyager I had the largest velocity, that is 16.5 km (10.3 miles) per second. Its engines did not have to generate this much extra velocity; the planets Jupiter and Saturn did most of the hard work. Presently, however, a new space machine is holding the record: 'New Horizons' is on its way to Pluto, with a velocity that topped 23 kilometers per second.

If we want to launch a spacecraft from Earth we need to use brute force, but once we are in interplanetary space we can use much more subtle means to move forward. This is only really true for longer voyages where the acceleration can be spread out over a period of time, and that's where possibilities might lie for alternative propulsion techniques. The ship's engine does not need to be as powerful as the engine that was needed to get us off the ground. It just needs to be very efficient. The point is this: the thrust produced by a spaceship's engine depends on two factors, the weight of the gases it blasts from its nozzles, multiplied by the velocity it manages to impart to these gases. Chemical forces cannot yield velocities much more than three km/sec. But the weight of the gases brought along in its fuel tanks also holds the velocity down. Therefore, we want to reduce that weight, which means we need to give higher velocities to the propellant gases than the chemical limit of three km/sec. This is where other energy sources should be useful. Nuclear energy would be perfectly suitable, but we could also use solar energy to accelerate the propellant gases to great velocities, a technology that has already been tried and tested during the European space project *Smart*, which is a small spaceship that encircled and explored the surface of the Moon.

Aside from solar energy, there are some other alternatives. It is conceivable that solar winds, a stream of extremely tenuous gases that is emitted by the Sun at high velocities, could be used to drive a spacecraft. But the gases are so feeble, that enormous sails made of extremely light material would be required. And even then the driving force is quite weak. I believe we might also be able to use magnetic fields generated by ionized particles emitted by the Sun. Long, extremely light cables made of superconducting material carry electrical currents, exerted by magnetic fields. In large loops, these cables wind themselves around the magnetic flux fields and pull one or two spacecraft along with them.

In the not too distant future, such technologies would enable us to travel to all the planets in our solar system within a reasonable period of time — months to maybe a year, or thereabouts. People are already researching and testing these techniques. Therefore, if you are fantasizing about colonizing Mars or the rings of Saturn (I will touch upon these later), then that is allowed as far as I am concerned; it falls within the boundaries I set earlier, or at least, the method of transportation is possible within the existing laws of physics.

I will talk about the future of our solar system in much more detail in a later section of this book. But will we be able to colonize the planetary systems of other stars? Will we be able to attain the immense velocity required to even get to these stars? This is highly unlikely. The star closest to us is Proxima Centauri. It is located at a distance of 4.3 light years. If we were able to travel at the speed of light, in other words at a speed of 300,000 kilometers per second, the journey would take 4.3 years. And the so-called 'relativistic effects' would occur: according to Einstein's relativity theory, the traveler would experience a smaller lapse in time than those who stayed behind.

Some speculate about the possibility of creating greater speeds for spacecraft by bombarding them from Earth with radiation, literally blowing a spaceship to its destination. But I don't see how that would work to any acceptable degree of efficiency. The error in this line of thought is that it is assumed that a rocket engine is terribly inefficient in its fuel consumption, because it accelerates not only itself but also its fuel; therefore, one argues that providing the acceleration using fuel on Earth should be much more economical. But both lines of reasoning are not true. Suppose that the emission velocity of gases used as propellant could be controlled (including the possibility of using the radiation of light as a propellant). It should then be theoretically possible to convert *all* consumed energy into propulsion of a useful mass! Think of it this way: by blowing its propellant backward, the spacecraft can in principle ensure that the final velocity of this propellant is much smaller than that of the vehicle itself. So, the mass that is used for propulsion is accelerated to the high velocity of the spacecraft only temporarily. None of the energy we used stays behind in the exhaust gases.

Now to use this observation fully would require that the emission velocity keeps up with the speed of the spacecraft, and this is technically difficult. If we are tied to one constant emission velocity, a small but technical calculation shows that we lose efficiency only by a factor two or three. That's not so bad. I have been asking myself what the optimal emission velocity might be if it has to be kept constant, and decided to try to calculate it. As it turns out, the optimal propellant emission velocity is 63% of the desired total difference in velocity. In this case the optimal energy efficiency of 65% is reached. This would indeed be very efficient, but in practice there would of course be various other technical limitations. What my arguments do show is that from the point of view of the use of energy, conventional rocket ships are not at all that bad.

Harvesting solar energy between the stars, far away from a sun — in interstellar space — will be difficult indeed. In between the stars, these suns shine only weakly, and this radiation is not nearly strong enough for the enormous amounts of energy required. The best method might be to position gigantic mirrors that reflect the weak sunlight, concentrating it to such an extent that it can be used as an energy source.

Nuclear energy is a more viable option. Nuclear powered rocket engines were designed, built and tested in the 1950s and 60s. In

June 1968 the last of the Kiwi series of nuclear rocket engines ran for over 12 minutes at a power level of 4,000 megawatts (the equivalent of four nuclear power plants!). Because of the problems associated with radioactive contamination, such engines could only be useful if used for massive spaceships operating far from the Earth.

But even then we wouldn't be able to come close to reaching the speed of light. The energy captured in the nucleus of the atom would be able to drive the nucleus forward at a speed of one-tenth of the speed of light, or thereabouts. So we would have the same problem as with chemical fuels: there is a limit to the exhaust velocity, which implies a limit to the final velocity that the vehicle can reach. And here too, we face the problem that the same engine would have to slow down the spacecraft at the end of the journey. That's why velocities of only a few percent of the speed of light will be the speed limit for any spaceship. Physicist and Nobel laureate Freeman Dyson has described a way to build a spacecraft that is driven forward by nuclear energy. By using a controlled fusion reactor that combusts frozen drops of hydrogen or deuterium, a spaceship would be driven forward at 1% of the speed of light.

The anticipated problems are huge — I see all sorts of serious constraints. With such high speeds, a collision with even the tiniest speck of dust would be fatal. In fact, we don't know the density of dust particles in or near our solar system and how this would restrict the permissible velocity of interstellar space travel. The limit might lie at a few thousand kilometers per second.

An important point to note in relation to space travel is that once great speeds have been attained, reducing speed is almost as difficult; there is nothing to push back against. To reduce speed, the engines need to be pointed forward instead of backward and the same amount of fuel is required as when the spacecraft took off. But is there really nothing to push against? Well, it might be possible to use the interstellar magnetic field. That field is extremely weak, but the superconductor kite rope I mentioned earlier may come in very handy. A loop of this rope, thousands of kilometers wide, would catch on to the weak magnetic field and decelerate our ship. As a free gift, an electrical current is generated, to be used on board as an energy source.

Anyway, we're digressing. The conclusion, then, is that a trip to Proxima Centauri will take more than a thousand years; not a very enticing prospect. I don't want to just look at the technical options we have — we also need to be realistic. Would there ever be a crew willing and able to make such a trip? They would have to pay their own energy bill, a bill for at least one large power station, which would have to run for many years. Of course, those left behind won't be willing to pay anything, because they won't get many benefits from this journey.

But the problems don't end there. How to survive the trip? Frozen maybe? There is a lot of speculation about freezing humans. It is not entirely inconceivable — more about this in Chapter 18 — but even if we were to accomplish it, what would we do upon arrival? Would we be able to grow a little vegetable garden? A chicken farm perhaps? Well, don't worry about the latter. If we ever manage to complete an interstellar voyage, then we won't find ourselves on completely new territory, quite the contrary: all the possible destinations within our solar system will have been thoroughly explored already and, wherever possible, plants and animals will have been introduced. How? With robots.

Robots can be very small, robust and are much better equipped to deal with the fireworks ignited by the many collisions with dust particles. They will be able to travel a whole lot faster than us. In any case, robots will precede humankind and I'm sure they can bring along a few seeds to start that vegetable garden. The advantages of using robots are evident. It will require, however, intelligent robots. I will elaborate on this in Chapter 13.

But how about those other suggestions, the space-time warp or whatever it is called, a voyage through a wormhole, a black hole or whatever? These ridiculous notions are forced into science fiction novels with astonishing ease and, as I pointed out in Chapter 1, even serious scientists are playing along with this nonsense. Usually they talk about a black hole, which according to Einstein's relativity theory would be able to make a connection with another black hole, creating shortcuts that bypass the fabric of our universe's space and time. This would be a handy tunnel, connecting points in different universes. Sort of like a galactic subway, really, and perfect for space travel.

But don't be fooled by these illusions. Let me be clear: black holes are very likely to exist, plenty of them are detected, even in our Milky Way. Rough calculations even appear to suggest that these are 'wormholes' to other universes or other parts of our universe. But a black hole is not easily manufactured; the laws of nature as we know them now, simply forbid it. What would be required to make a big enough black hole is in no way compatible with the laws of the so-called Standard Model of Elementary Particles, a model from which no deviations have ever been discovered, not even with the most powerful particle accelerators and sensitive detectors.[d]

And even IF it were possible to create such a black hole, then the gravitational force effects alone would disrupt the entire solar system. To create a hole of only three meters in diameter, a mass as big as Jupiter would be needed. That hole would still be much too small; anyone coming anywhere near it would be ripped apart immediately. Only black holes that are thousands of times heavier than our Sun would be large enough for a human space traveler to avoid being torn to pieces by its gravitational tidal forces. And you would want to make sure you didn't encounter any other matter while you're in a black hole, because it would have turned into deadly radiation against which the reckless space traveler wouldn't stand a chance. Astronomers have seen enormous black holes in the Milky Way and other galaxies, but they look decidedly deadly!

---

[d] The recent experimental detection of neutrino mass is easily incorporated in the Standard Model and doesn't change a thing regarding the above.

Even if we imagine that the space traveler were somehow able to protect himself against the radiation, then it would still only be a black hole and not a wormhole with a nice trapdoor to the other side of the universe, because that door would have to be made of matter that has characteristics which are impossible according to modern physics. And even if the required matter existed, then there wouldn't be any way to construct such a door. And even if... then... well, you get my drift.

I would have stopped after the very first "even if, then...", were it not the case that 'warps' pop up everywhere and even Stephen Hawking and Lawrence Krauss devote serious discussions to this subject. Space warps, of whatever similarly vague principle, are entertaining for science fiction novels, but deserve no place in serious physics and should not be confused with a realistic outlook on the future.

So back to reality: there is an alternative to nuclear energy that should not be ignored. In principle — and we are not looking at the technical obstacles for a minute — there is the following possibility: assume that large amounts of *antimatter* can be manufactured. It would cost tremendous amounts of energy, but that is unavoidable because traveling to the stars demands high velocities that require whopping amounts of energy whatever the method. And then imagine two spacecraft, one made of matter and one of antimatter. As you may know, as soon as matter is brought into contact with antimatter, an annihilation process takes place that converts most of the total mass straight into usable energy.

So now, both spacecraft send a beam of their matter to the other, ensuring that the annihilation process occurs at exactly the right time at the right spot. This creates huge temperatures and pressures, as well as exhaust velocities nearing the speed of light. How we could create a ship of antimatter, I have no idea; unfortunately, it is not possible to store antimatter because any vessel made of ordinary matter would explode immediately when coming into contact with it. And all the tools we need to build the ship would also have to be made of antimatter, handled by antimatter robots, and nothing made of antimatter can come into contact with anything made of ordinary matter, including the air. I am afraid this plan will remain unrealizable forever.ᵉ

Even though the practical difficulties with such a method of propulsion would be incalculable, it means that boundaries such as those associated with nuclear energy are, at least formally, not set in stone. However, it is most unlikely that traveling to the stars will be feasible other than through conventional methods, which means using engines that emit gases to create forward propulsion. It also means that a voyage to even the closest stars will take thousands of years.

This appears to be quite unacceptable for most science fiction writers, and is even ridiculed. This is mainly because of a lack of understanding of the level of scientific knowledge we have been able to accumulate over the last two hundred years, particularly in the area of physics. While it may be true that radical new developments in physics are not entirely out of the question, what we already know in this area makes it extremely unlikely that humans of flesh and blood will ever make the trip to another star in our universe. It is important to realize that I am not just talking of human limitations, but also of those imposed on aliens who so many people seem to think are on their way to meet us. Even if they were to be millions of years ahead of us in terms of technological skills, the laws of physics would apply also to them, and it would be highly unlikely that they would be able to traverse the enormous cosmic distances much faster than us with the methods I just described.

It should be clear that the price to be paid for a one-way trip to Sirius and the expected sacrifices and deprivations to be made because of the unfathomably long duration of the trip, probably suggest that such trips will never be undertaken by humans. Such

---

ᵉAnything made of antimatter would have to be constructed anti-atom by anti-atom. It was once thought that objects made of antimatter may exist elsewhere in the universe. Nowadays this is thought to be extremely unlikely.

a trip would not add any value either, because robots could go for a fraction of the cost and wouldn't care if the journey took tens of thousands of years. Plus, they would be able to explore new planets just as well as we could. Will robots be sent into space to make such voyages? And if yes, why? Please continue to read.

For the next millennia, the icy dwarf planets beyond Pluto's orbit will be the outer limits of where we humans will be able to get to. Would we be able to live on these freezing worlds? Definitely. Would there be anything of interest there? Perhaps. There are various exciting prospects for colonizing our solar system.

chapter 11

The Colonists

It was July, 1965. I was at my parents' summer house in the south of France, far removed from civilization. But the newspaper, which I loved reading, back then as much as now, was nonetheless delivered to the cabin every morning. I expected some news about planet Mars — and I got it: the first American spacecraft to visit the red planet, the Mariner 4, had arrived and was sending pictures of the surface to NASA's space centre on Earth, the Jet Propulsion Laboratory (JPL).

An earlier American mission, and various Russian attempts, to send unmanned spacecraft to Mars had failed. The Mariner 4 took a total of 22 pictures. Two were printed in the newspaper I was holding. They were unfocused, but you could distinguish the planet's craters.

The American space program had really taken off. In July 1969, the crew of Apollo 11 landed successfully on the Moon's *Mare Tranquillitatis* and returned home safely with pockets filled with Moon rocks. My fantasies of human colonization of the Moon and other planets had now become unstoppable.

How will this develop? Will there be human settlements, first on the Moon, and then on Mars, the moons of bigger planets like Jupiter and Saturn, and even on asteroids? Is Mercury inhabitable?

The large gas planets are unapproachable for humans. Their atmospheres are too heavy, their gravitational pull is too hard and they don't have a solid surface. Their moons, on the contrary, have little gravitational pull and most of them do not have an atmosphere of any significance.

In recent times, many have speculated about 'terraforming', particularly with regard to Mars. Human colonists would try to create an inhabitable atmosphere and raise the planet's temperature to acceptable levels — Mars is extremely cold. We now know that there is water on mars — in the form of ice. Some ice layers have been measured to be more than several kilometers thick! Yet this may be far from sufficient for what is needed for terraforming. Water and air would probably have to be imported from somewhere, maybe by dropping one or more small icy asteroids onto Mars (I've got more to say about influencing small planets' trajectories later on in the book). A lot of carbon dioxide would be required to generate greenhouse gases on Mars, which receives a lot less sunlight than Earth because of its greater distance from the Sun.

I promised to discuss science fiction from a scientific perspective and to leave other considerations such as practicalities and viability till later. Alright then. The laws of nature do not forbid terraforming from a theoretical perspective, but it doesn't appear to be a very realistic possibility. For Mars, influencing its climate is imaginable in the long-distant future, but terraforming on the Moon is not at all an option. Any manmade atmosphere would be too unstable on the Moon; because of the weak gravitational pull, the air would drift off into interplanetary space.

The latter is a bigger problem than you might think. It would take a long time, but maybe within a few hundred years the dissipated air would eventually return to Earth because of our planet's gravitational pull. There, it would cause an unwanted contamination. It would be much more realistic to create enormous domes, made of strong glass, underneath which an atmosphere could be conducive to humans. The glass industry would have to go all out, but the raw materials to create these huge glass bubbles seem to be available on the Moon.

However, there is another problem we need to face: radiation. A lot of dangerous and harmful radiation is present in the enormous empty space around us. At home, the Earth's atmosphere protects us. Glass domes would only be able to protect us if the glass were thick enough. The temperatures of the frozen moons of the gas giants are so low, we could even use the ice found there to protect us. Ice is a marvelous building material, as anyone who has ever tried to make a snowman or an igloo, will know. The glass, or ice for that matter, would have to be very thick, particularly if we want to build huge domes, because the larger the dome, the larger the atmospheric pressure underneath it. We would have to strengthen the domes by adding rocks or, even better, using layers of stronger, transparent material.

The first colonists on the new worlds would not have glass domes to live under. I think that the first dwellings would have to be underground. No radiation, easy to create a pressure of one atmosphere, but unfortunately, disappointing views. The colonists would have to build their own, first, domes.

Cutting to the chase, which bodies in the sky are we talking about? Many would be suitable. The planets Mercury and Mars, the Earth's Moon, the larger moons of Jupiter (Io, Europa, Ganymedes and Callisto), the moons of Saturn (Mimas, Enceladus, Tethys, Dione, Rhea and Iapetus), the moons of Uranus (Ariel, Miranda, Umbriel, Titania and Oberon), the moon of Neptune (Triton), our dwarf planet Pluto and its moon Charon, asteroids such as Ceres, Vesta, Pallas, and many other smaller moons and asteroids in interplanetary space.

Not all bodies are equally suitable. Mercury, for example, is quite close to the Sun, making the daylight temperatures intolerable. For a long time, it was believed that Mercury always turned the same side towards the Sun, so that one side would remain cool, but that turned out to be false. As a result of the myriad gravitational pulls by the Sun and the Earth, Mercury shows the same side whenever it is visible from our planet. But Mercury does rotate, even if one day is 176 times longer than a day on Earth. Nonetheless, I mentioned Mercury because it would be technically possible to create cooler temperatures there, as there is no atmosphere.

Other choices, such as Io, receive so much radiation, that we would have to dig incredibly deep to protect ourselves. Furthermore, this world is one big volcano. In addition to the fact that its sedimentation contains a lot of sulfur, this means that people would never really feel very comfortable here. And many of the ice worlds I mentioned are so far removed from the Sun, that it won't just be extremely cold — which by itself is not really a problem — but there is little solar energy to be had.

In 1974, an article written by Gerard K. O'Neill appeared in *Physics Today*, a professional journal, which appealed to the imagination of many. According to O'Neill, we should be able to create our own worlds in space, in the form of big cylinders that revolve around their axes. The first cylinder would have a diameter of about two kilometers and would be roughly five kilometers long — picture a beer can, rotating around its longest axis. As a result of the rotational motion, the inhabitants are pressed against the curved inside of the cylinder, a bit like laundry in a washing machine, a force that mimics a gravitational pull as on Earth. Sunlight enters through three large windows. There is air and water, agriculture and animal husbandry, and the inhabitants are completely self-sufficient: these would be true space colonies, where Earth-like climates are created. The first cylinders would move in orbits between the Earth and the Moon, but later versions could be everywhere in interplanetary space. The building blocks are not from Earth — that would be much too costly — but from the Moon, mini planets or nearby asteroids. Every cylinder would be able to accommodate tens of thousands or perhaps even millions of people.

One of my colleagues claimed this would never work; he argued that no material exists that would be able to withstand the pressure from the rotational movement and from the air on the inside. But

Panorama of the interior of Gerard O'Neill's space colony. Considering the building material currently known to humankind, the cylinder would have to be a great deal smaller.

I don't agree. The casing would have to be meters thick, but it could definitely work.

The theoretical limit of one of these cylinders has actually been calculated; apparently, it could reach 19 kilometers long! I am not sure though whether those who carried out the calculation took into account the problems concerning the supply of light and the draining of excess heat. The walls would have to be very thick — at least five meters or so, and made of the strongest type of steel — and this could be problematic. I would limit its size to a few kilometers at best. Only if we accept a slightly lower gravity and air pressure than the ones we are used to on Earth, could we increase its size.

Everything inside the cylinder would be recycled, so water, air and other necessities would only have to be imported once. Almost all the raw materials would come not from Earth, as transportation would be much too expensive, but from elsewhere. Most necessities would come from the Moon, except for water and hydrogen. These we could retrieve from asteroids, in the form of ice; there is lots of ice to be found on asteroids.

It is remarkable that the enormous amounts of the strongest steel required to build such an inhabited space cylinder would actually be available. Some asteroids consist mostly of iron and nickel. The largest such asteroid is Psyche, more than 25 kilometers wide, but she is very far away. Smaller iron and nickel asteroids are actually ubiquitous; we would be spoiled for choice. The smelting would have to be carried out in ovens in space, but whether this is technically possible, only the future will tell.

A huge advantage of these colonies is that the inhabitants would experience gravity similar to that on Earth, which would help them to feel more at home. A likely disadvantage is that the large funds required to build such a cylinder would be difficult to raise. Perhaps a chosen few pioneers could first make a small asteroid inhabitable by creating a cave, from where they loot nearby asteroids to create a better habitat until, eventually, something resembling an 'O'Neill Wheel' emerges.

It is more likely, I think, that future colonists will have to make do with less or even without any gravity. Their bodies will have to adapt, but that is probably easier than building huge cylindrical habitats. The most exciting future scenario would of course be one where all sorts of different colonies coexist peacefully, on, in and orbiting various planets.

Calculations show that space colonies would be able to orbit planets far beyond icy Pluto, as long as we've got large mirrors that catch all the available sunlight, heat and energy. There might even be daredevils who choose a hyperbolic orbit, on course to another star; but because of its size, they wouldn't have the velocities needed to reach their destination much earlier than a couple of million years.

What I have described here, in summary, is a dream world confabulated in the '70s. However, I think there are various issues that were not considered back then, which could give an interesting spin on things in the future. Let me tell you what I mean in the next few chapters.

chapter 12

The Cambots

MANY OBSERVERS HAVE noted that in the last few decades all sorts of developments have taken place that most science fiction authors did not see coming at all. One example is the Internet. There were a few though, who did. I remember a story, long ago, of a journalist who did not only report the news, but who also reworked and reported the information he received in such a way that he completely controlled what happened next, literally making the news, without anyone noticing. As it happened, he was extremely well-connected to all the computers in the world and he possessed a world class computer himself, one that was semi-intelligent.

Most stories of space travel, however, do not even mention the Internet. Sci-fi writers failed to realize that, with every journey of any consequence, certainly with every important trip into space, half the world would be following the developments closely, glued to their television or computer screens. Clearly, whatever else the information revolution will bring us will be of decisive importance for the future of the conquest of space by humanity.

I have already discussed robots. Look at the state of the world today: robots are the most efficient creatures exploring planets and their moons. While I write this, two unmanned vehicles, called

Opportunity and Spirit, are roaming on Mars. They send us images, conduct analyses and are completely remote-controlled. Much further away, a spacecraft named Cassini is in orbit around Saturn, and creates numerous incredibly clear images of the planet, its rings and the surrounding moons.

In the meantime, a probe named after the Dutch scientist Christiaan Huygens, has detached itself from Cassini and has managed to land independently on Titan, the largest of Saturn's moons. But you probably know all this already, as you will have followed it all closely via the Internet and other media. As it appears, it is much easier to explore space using robots than manned spaceships. Not only will this continue to be the case, but the advantages that robots have over man will increase, and quickly too.

Why was it again we wanted to go to the Moon? Ah yes, to see what it is like up there. Robots can tell us that much better and cheaper than if we were to go ourselves. They can travel, adapt to nearly any kind of environment, make observations in all colors of the spectrum, listen and measure. Why should we humans go at all?

Let's not beat around the bush, there is only one real answer to that question: because of expansion drift. We want to be there to conquer, to take charge of those faraway worlds, and to experience for ourselves what it is like to be there and to soak up the atmosphere. We want to because we can, even if the costs are sky high. We want to establish Earthly, biological life — human life in particular — on other planets and in space itself. Perhaps other motives will emerge for us to want to set ourselves up physically in faraway places, such as for political or juridical reasons. Or perhaps because of proprietary rights. And then we have military demands. Quite conceivably, the military would insist on human observation posts in space.

In the meantime, robots carry out those tasks one would expect from humans, and they are better at it. However, our current robots are not intelligent, which makes them slow and vulnerable. The

Mars vehicles Opportunity and Spirit are only able to move very slowly, because the drivers are on Earth. Depending on their relative position, signals back and forth take anywhere between eight and 40 minutes, and thus the same amount of time passes before we can observe the reactions to our steering commands.

A lot can, and does, go wrong. Nevertheless, we have made a lot of progress already and all sorts of small defects can be repaired. Our current capabilities in the area of electronic intelligence are thoroughly tested by how we handle these situations, but this is also where spectacular improvements are imaginable. Future missions will be accompanied by much more intelligent computers and it might even be that the first pioneers setting up colonies in distant locations will be robots, not humans.

But aren't there plans already to send people to Mars? Well, they might stay for a year or so, but then, sadly, they'll have to return home. What I am afraid of, is that these sorts of undertakings will end up just like that splendid American Moon program. No one has visited the Moon in the last three decades. There's no one there now, and the enthusiasm to colonize the Moon on a more permanent basis has dwindled. The voyage to and the sojourn on Mars will be much more expensive than the Moon program. Maybe the first mission will be followed by a second one, perhaps even a third, but then money and excitement will certainly run out, and there will be no further interest in permanent colonization. Just like with the Moon, our expansive urges will have abated.

I am afraid that permanent colonization of Mars is nowhere near becoming reality. Instead of a premature human visit to Mars, it would be better to first establish whether a human colony in space is even possible. The Moon is the ultimate guinea pig. Shouldn't we try to establish a permanent colony on the Moon first? Not only would we gain invaluable experience, but above all, a launch from the Moon requires a lot less energy than a launch from Earth — a factor of 20 less! Thus, future Moon dwellers would be in a much better position than Earthlings, or those from whatever other celestial body, to successfully colonize Mars.

How would the first colonists survive on the Moon? Let's assume that not only will there be enough volunteers, but also a willingness by those left behind to pay for the extremely expensive trip and stay, because it will take a substantial amount of time before the pioneers will *really* be self-sufficient. But that is what true colonization is about.

How will these colonists acquire the energy they need, as well as the air, water, food and security? How will they expand their new home when the colony starts to grow? Is this really imaginable? This is where that new invention called the Internet, which most science fiction writers didn't foresee, should be exploited as much as possible. There must be loads of possibilities for that.

Every now and then it is speculated that future Moon colonies will survive because of tourism. Tours will be organized to a Moon hotel, where vacationers can play sports, wander around and make excursions. Or perhaps people with particular medical indications will make the trip to the Moon, where gravity is much less than on Earth. Those people will have to bring in the money required for the colony to expand.

The project I am about to propose will be able to sustain itself much sooner. It will be possible by offering all those people back on Earth who are using the Internet something of value. A space travel organization like NASA or the ESA — though a private organization with as much government funding as it can get might also work — will launch a couple of one-way trips to the Moon. The most important passengers on board won't be humans, but robots. Not intelligent robots, we can't make those yet, but smaller versions of Opportunity and Spirit; small vehicles, in other words, decked out with hands and tools. Each will be equipped with a camera, a fancy webcam of sorts, and that's why I propose to call them *cambots*. In addition to these 'passengers', the first mission will also carry an electricity generator, which will have to last quite a while — perhaps even a small nuclear reactor. This will provide the necessary fuel to the robots. Further, we will need a small excavating machine, for reasons that will become apparent shortly, and an extremely strong communication channel.

The required technical expertise for the above already exists and has been thoroughly tested, so nothing I have suggested so far cannot be realized within a short time frame. But now the new thing: the cambots will be *leased* to Internet users on Earth. Users will be able to rent a cambot for half an hour, a day or a month, as long as they want — dependent only on their financial resources, because it won't be cheap. Further, a hefty deposit will have to be made to prevent the robots from purposely or accidentally bumping into things, damaging themselves or other objects. To further prevent irreparable damage to the cambots, software will be introduced, such as simulation programs and joysticks, which will enable users to practice controlling a cambot. This is highly recommended, and maybe taking driving lessons and a test will be mandatory before anyone is allowed to lease one. It is not at all easy to control a cambot from your lazy chair in front of your computer, because the signal to the Moon and back will take approximately three seconds. The effects of your commands will only be seen three seconds after the command has been given.

A computer screen will display the images made by the cambot and show the robot's hands. It will be driven all along the lunar terrain, where it will pick up rocks, though the driver will have to ensure that the cambot's battery is charged. Of course, it could also be possible to lease a cambot that is connected to the main energy provider by a wire, in which case the driver would not have to worry about recharging.

Cambots will have their own territory on the Moon, dependent on their action radius — maybe only a hundred meters to begin with — and all sorts of games between cambots could be invented. Cambot handball perhaps. While it may not be as exciting as Harry Potter's quidditch, it will require a lot of dexterity by the players.

These sorts of strolls and games with cambots will become boring quickly, if the cambot drivers are not challenged by specific tasks. After all, we want to expand and perfect our cambot colony. First we will build roads for the cambots, to enable future cambots to take comfortable strolls across the Moon. Charging stations will have to be built at the road sides and industrial activity will need to be set up. We want to harvest and process raw material and for this we need factories. In particular we will need to create a glass factory to manufacture windows for our dwellings, a larger energy provider, and in the longer term, a cambot factory. The ultimate goal, of course, is that the area where the first cambots live becomes suitable for human occupation. In other words, it will be the cambots' task to create a comfortable lunar hotel.

Consequently, cambot drivers will be used as employees and they will want to be paid for their services. Those who are good at driving their cambots may be able to earn back their initial lease investment, or perhaps even make money out of it. Maybe some cambots will have to make do with credits that are paid out only when the first guests of the lunar hotel arrive, but I am not sure what financial arrangements will be put in place. I don't know much about finance. A lot will depend on the enthusiasm and effort made by the cambot drivers. What gives me hope is that a venture like this will be less dependent on politicians and bureaucracy and more on the effort and creativity of the general public. Further, there are organizations such as The Planetary Society, a mighty American organization of space enthusiasts, who have a lot of money. I think they would heartily support such a venture and they would want to take an active planning role.

As soon as there is enough glass to create airtight structures, seeds, chickens, eggs and fish will be imported. Thus will be the beginning of our lunar farming operation. There is an important point to note here: cattle farming and horticulture will not be easy, and will require a lot of expertise. Internet users from all over the world will undoubtedly possess enormous amounts of knowledge and inventiveness. I am counting on it. The first steps will surely consist of trial and error, but human ingenuity will prevail. I am thinking about the biochemists who planted a rainforest in the savanna of Colombia. Their great accomplishments may well be dwarfed by what is required to make agriculture and farming a success on the Moon.

I am counting on the Internet to provide the necessary solutions. The cambot drivers are continuously in contact with each other and ask third parties for help. The numerous sites and chat rooms will become very popular and useful, but there will also be countless charlatans who will want to contribute, people with wild and unrealistic ideas. The serious workers will have to be able to make use of secure communication channels, and the most important knowledge databases and websites will have to be maintained by experts.

But why shouldn't we allow a professional space organization to manage this operation? Why are we letting an inexperienced public push the buttons, with all the associated risks? The reason why I hold on to this vision, despite these objections, is because it is the best way to ensure continuous global political support for our project. Sure, things will go wrong, but if reparations are required, there will be sufficient funds to cover them. Perhaps this will be the way to have private enterprise do the jobs that until now were the privileges of government agencies.

To my delight, I heard that such private initiatives already exist. 'Astrobotic' is a company that designs and manufactures robots. They also aim to send their robots to the Moon. Google has issued a $30m prize, the 'Lunar X Prize', for the first who can drive a robot over 500 meters on the Moon. Astrobotic plans to collect that prize and continue from there. Their robot will revisit the Apollo 11 landing site. The public can follow this 'Tranquility Trek' in high-definition video, "experiencing the lunar adventure with the same clarity as Neil Armstrong and Buzz Aldrin," according to the promises in their brochure.

I also heard of another interesting source of income. The cost of sending material from the Earth to the Moon will be a few million dollars per kilogram. This means that, perhaps for as little as $6000 or so, it can be arranged that a few grams of the ashes of a deceased loved one can be sprinkled on the Moon. The sprinkling can be followed in high-definition video back home. This is just one of those imaginative ideas that may help such endeavors to become a success.

If things go the way I think they might, then many television stations and other media will want a piece of the pie. Possibly, rich industrialists will be prepared to sponsor cambots and other elements of the project. And the more that is achieved, the more money will be spent and the easier it will be to keep the whole project afloat financially. I fear that the first lunar hotel guests, but also many guests after that, will have to concede much of their privacy; it's for the greater good, but a sacrifice none the less. Accordingly, I have started to watch reality shows such as *Big Brother* with new eyes.

The first lunar hotel will be mostly underground, for both practical and security reasons. The first guests will get bored quickly and will want to go back home, but as the colony and its activities grow, they will want to stay much longer. The facilities will improve along the way and because gravity on the Moon is lower than on Earth, fantastic sports games will be invented. It will be easier to control the cambots from the lunar hotel than all the way from Earth, because there won't be a three second delay with every command. So, as more people move to the Moon, they will slowly but surely win the competition with the cambot drivers on Earth — or at least, that is what is supposed to happen. The lunar colony will then be able to develop its own independence. It will undoubtedly take years before we reach that point, but once started, the colony will spread over the Moon like a rash.

Once our lunar colony has established itself successfully, we could turn our attention back to Mars. Now that we have obtained loads of experience colonizing the Moon, we could attempt the same thing on Mars. However, this will be quite a bit more difficult. The delay between a command and the cambot's reaction will be 20 minutes or so on average. We will have to find a solution for this.

A hotel on the Moon. It boasts to be the only place on the Moon with Earth gravity. With a diameter of 125 meters, it makes four turns around its axis in one minute. Walking on the inside of the parabolic wall feels like standing up vertically. Further from the axis, gravity increases, until it reaches the same strength as on Earth. It will be quite comfortable, albeit a bit unfamiliar. The smaller tents in the foreground of the picture were built by earlier settlers.

Well, our information technology is continuing to make steady progress and all sorts of handy software assists us to ease the effects of the communication delay. And our cambots are getting more intelligent: while they might not yet have human ingenuity, they are more and more independent. A cambot that falls over does not have to wait for commands from Earth to get back up again on its wheels — or feet? — and they will have the ability to locate their recharging sockets on their own. And thus the cambots commence their work on Mars.

We won't see visitors to Mars anytime soon. What we don't know yet is what the first permanent lunar colonists will have to tell us about their emigration experiences to the Moon. Or whether potential Mars adventurers would be inspired to actually risk that one-way trip to the red planet — or rather book a (much more expensive) return ticket. The fact remains that traveling to Mars from the Moon will be much easier and cheaper because of the lower gravitational pull of the Moon, which also means that lunar colonists will take an integral part in colonizing Mars.

I hope that, because of the central role of the cambots in conquering space as I envision it, toy manufacturers will introduce simple versions of the cambots to the market place. Of course, these toy cambots will never see the Moon up close, but we can certainly practice controlling them. And we can start drafting those rules for cambot handball!

chapter 13

The Neumannbots

THE IDEA OF creating reproducing robots is an old one. Just before he died in 1957, mathematician John von Neumann wrote a discourse in which he provided what he claimed to be the mathematical proof that self-reproducing robots can exist.

Well, based on our current knowledge, it is easily shown that such robots do indeed exist; every living organism is testament to that, after all, they reproduce themselves. It is one of the most amazing features of our universe, that after a long series of extremely complex processes, creatures emerged that are able to produce offspring. But that's not all. Not only do they have the capacity to reproduce, but they can also make small changes in these reproductions. We call these changes 'mutations', and in the long history of life on Earth, these mutations allowed progeny to adapt to varying circumstances and, eventually, to supersede their parents.

These brilliant examples of terrestrial life are what we have in mind when talking about reproductive robots, but life also shows us that there are certain inherent dangers. The self-reproducing robots we call 'life' are programmed in such a way that they are firmly determined to reproduce themselves at all costs. This causes severe competition between robots and every now and then will generate tough and unpleasant battles. It is certainly not in our best interests to create a new, more aggressive life form, which could be more sly than humanity itself, posing a threat to our own existence. Building robots such as those envisioned by von Neumann would not be without risks. Science fiction authors know precisely how to write exciting stories based on this theme.

Let us ask biologists to explain why living and reproducing creatures are continuously in a life and death struggle with other creatures. It appears that this has everything to do with the way in which information regarding reproduction is passed from individual to individual: through their genes. Every individual carries a complete set of genes and those genes determine how creatures reproduce.

Richard Dawkins knows everything about this and he has elaborated his views in his book *The Selfish Gene*. Genes are parts of chromosomes, which are multiplied during reproduction. The 'more valuable' the information that is stored on a given gene, the larger the number of these genes that are successfully reproduced. This is a biological law and it holds true whatever actual behavior or characteristic is contained in or triggered by the gene. The key message in his book is that the behavior of living creatures is solely a result of the way in which genetic information is passed on from parents to offspring. If we were to allow our robots to pass on genetic information in a similar manner while reproducing, then after a long period of evolution they would begin to obey the same biological rules, and behave somewhat like human beings.

Clearly, we do not wish our robots to become violent. If we handle it well we can prevent this from happening. Those who read Dawkins' books carefully may have learned what needs to be done to ensure this. We should never allow ourselves to deck out our robots with mutating genes; that could prove too dangerous. The robots should be controlled by external computers, which are in continuous contact with each other and exchange information. If we monitor the computers closely to ensure that they continue to exchange information, then the robots would behave like ants in an anthill. They would all be biologically related and only interested in co-operation, completely in accordance with Dawkins's theories.

In any case, let's continue to call the self-reproducing robots 'neumannbots'. Although the neumannbots are intelligent, their processing information is still stored in large central computers. Eventually, the neumannbots would replace cambots when colonizing interplanetary space. Through our cambot exploitation, we would have learned how essential tasks are to be performed, but unlike the cambots, the neumannbots would be able to carry out these chores by themselves. An advantage of Neumannbots is that whatever useful information is learned by one can be instantly passed on to the others. This, however, should never happen without proper authorization from the main control center on Earth, because of the imminent dangers of 'genes running wild'.

Currently, two immense technical requirements have not yet been met, and they are standing in the way of the neumannbots' development:

- The need for far-reaching artificial intelligence.
  It is certainly not possible to produce such intelligence with just a snap of the fingers. The cambots will develop and improve themselves, but the first models will not be anywhere near intelligent enough to be able to work independently. Our army of cambots will have to go through a lengthy learning process, but as they gain more experience they will become more independent.
- The ability to produce all required parts themselves, given the fact that they will have to identify, collect and convert all raw materials.
  The delicate electronic components — chips and the detection methods based on advanced nanotechnology — will have to be manufactured by humans on Earth for the foreseeable future, because I don't think it is likely that the small robots will master the incredibly complicated techniques required to create electronic components any time soon. It is conceivable however that Neumannbots will eventually succeed in making

integrated circuits and other electronic components, using highly miniaturized special purpose packages.

The neumannbots would be able to build up colonies much faster than the cambots, which are dependent on commands from their human masters. And because they are not hindered by any delays caused by the distance between Earth and wherever the cambots are, neumannbots can be sent much further into space.

I doubt whether it will be worth our effort to establish accommodation for homo sapiens in every colony we set up. Cambots and neumannbots will be able to reach places that are simply inaccessible to humans: too hot (Mercury, Venus), too cold (Titan) or too dangerous (Io). There might also be radiation, or there are other inhibiting factors. Specially-adapted neumannbots would be able to withstand those environments where humans do not stand a chance.

And then there is something else: the robots of the future — they don't all have to be neumannbots — will be able to send images and impressions from their habitat to people, with much more detail than our current primitive spacecraft will ever be able to do. The information technology will have been revolutionized by that time to such an extent that enormous amounts of detailed images and other essential information will come back to us on Earth. During virtual reality sessions, we will be able to experience the robots' adventures from our own comfortable and safe living environment.

The vast majority of humanity will not be able to experience their space adventures in any other way, and I think that most will be quite happy where we are: virtually positioned on a robot's shoulder we'll feel what they feel and without too much effort imagine ourselves far into interplanetary space. Whatever the robots see is real and therefore what we see is real: we'll revel in it.

I have mentioned all the planets, moons and asteroids where neumannbots would be able to set up colonies and many of these planets will be accessible for humans. However, I have not yet mentioned one very special location. While I write this, the

American robot Cassini is traveling in a beautiful voyage circling around the magnificent planet called Saturn. In its vicinity there are dozens of small moons looking perfectly lovely and inviting. But what about those rings?

Saturn's rings consist of rocks of various dimensions, varying between fine gravel and boulders of a few meters in diameter. They consist mainly of ice and are arranged in beautiful patterns. Their origin must be sought in one or more smaller celestial bodies that ventured too close to the huge planet, which caused them to be ripped apart by its strong gravitational pull. As a result of collisions between the rocks, some were pulverized and, under the influence of other moons and the planet itself, the orbits along which the debris settled themselves were divided into 'allowed' and 'disallowed' streams around the planet.

The 'allowed' orbits form the stunning patterns of circles we now see. The darker areas between the circles are the 'disallowed' orbits, where hardly any debris is found. Cassini shows that the former are formations of only a few kilometers wide. This means that the rocks do not want to be separated from their ideal position by more than a few kilometers. Some scientists even estimate that these rings are only a few meters thick. The time it takes to go full circle around the planet varies between a few hours and half a day. The fact that they all stay so close to their ideal orbits during each revolution must mean that the rocks collide with only marginal relative velocities, perhaps no more than a few meters per hour. That is very slow! A spaceship could easily navigate between these rocks and a collision would not cause any structural damage — it might even be able to establish a colony!

Let's start with a robot colony. The major advantage of a location between Saturn's rings is the absence of gravity there, but also the abundance of raw material to use for construction. I am not sure about the radiation though. I remember reading somewhere that there are very few ionized particles, because the material making up the rings absorbs them so well, but I could not get this confirmed. In any case, hard UV and X-rays radiated by the Sun will certainly cause harm.

The rings form a surface of about hundred times that of Earth (science fiction authors would be able to conjure up marvelous stories of super-powerful kingdoms controlling the entire galaxy from here). On the other hand, the trip to and from Saturn's rings would cost a lot of energy because of the planet's strong gravitational field, whose forces would have to be overcome. But consider the option of placing fusion reactors there, after all we have more than enough hydrogen. Energy will never be a problem!

Life on Saturn's rings would indeed be unique. Every ring moves with a slightly different velocity around the mother planet, so that sitting on one ring, all the surrounding rocks would appear immobile. However, rocks on a nearby ring, let's say a hundred meters removed, would appear to be moving slowly with a speed of a few dozen meters per hour. Nearby rings won't differ much in terms of their velocity.

At times, storms erupt, when the moons are positioned just right or because of external intrusions, for example a comet. Then the rocks are shaken about a bit more violently. In most of the rings, the storms are inconsequential because neighboring rings absorb the shock wave, but the outermost ring, the F-Ring, looks somewhat more volatile. Here, the moons cause a bit more of a stir and the rocks are further apart, having less of a tempering effect. In other places, motion is completely constrained. When perfectly aligned, these boulders exude an image of being at peace with the world.

In my vision, Saturn's rings are the perfect breeding ground for neumannbots. Raw material is there for the taking, no complicated maneuvering mechanisms are required, and it is possible to execute large scale expansion, enabling each robot to concentrate on carrying out its own specialized tasks. Subsequently, some of the neumannbots could disperse towards the outer rings, towards the Saturn moons, and back into interplanetary space, where larger projects await. It should be noted that the other gas planets such as Jupiter, Uranus and Neptune also sport rings, but they are not as prominent as Saturn's.

The neumannbots would not suffer as much from the cold and the lack of energy as humans, and would therefore be able to travel much further away from the Sun than human colonists. Out past Pluto, there are several other dwarf planets, separated over immense distances. A few of them have since been identified, but astronomers have deduced from their observations that there must be thousands of planets and asteroids, and millions of smaller objects, such as comets.

These icy rocks form the so-called Kuiper Belt, a large disc around the Sun, located beyond Neptune, where they move in almost circular orbits. The belt is named after Dutch astronomer Gerard P. Kuiper. The mini planets move in twilight and it is extremely difficult to detect them. Astronomers possess detection techniques that are no way near reaching their theoretical limit. They will discover loads more things in the future, either with new generations of telescopes or with advanced radar systems, meaning that the future destinations of even the most travel obsessed neumannbots will be known. There are no limits, but their travels will take longer and longer; first ten years, then hundreds. It doesn't matter, the neumannbots have all the time in the world.

There are millions of small worlds, out there in the Kuiper Belt. How many of those will be visited by neumannbots and eventually inhabited by humans? And will the robots be tempted to explore the galaxy even further? Beyond the Kuiper Belt, we have the Oort Cloud, named after its discoverer, the Dutch astronomer Jan Hendrick Oort.

The Oort Cloud is a more diffuse collection of icy objects. As stars cause the cloud's orbits to be imperfect ellipses, every now and then an object is pulled from its orbit by a neighboring star. So it can happen that an ice ball sways into an orbit that brings it closer to the Sun, sometimes even close to Earth. If that were to happen to an object, which hasn't been anywhere near Earth in billions of years, it is subjected for the first time to the vicious radiation of the Sun. Frozen gases and water evaporate and become ionized. Small dust particles are blown away, taking the appearance of a tail. This is when we see a comet in the sky. Oort estimated the number of icy objects based on the amount of comets we see each century. He concluded there must be billions of them, just snow balls really, some hundreds of meters across.

Unlike the Kuiper Belt, which looks like a disc, the Oort Cloud resembles a more spherical cloud around the Sun. The icy objects in the Oort Cloud are scarcer and spaced even further apart than those in the Kuiper Belt, probably bordering other Oort Clouds surrounding other stars. The neumannbots will reach them nonetheless. It will take thousands of years to do so, but then our neumannbots will have finally established their presence in interstellar space! Human colonies will be out of the question, but not human curiosity.

Economically, such an expansion will probably not make much sense. As long as human colonies are set up, there is likely to be trade: human colonies on the Moon and Mars will have to satisfy their never ending quest for water to quench their thirst, and water is available in abundance at colonies further afield. The faraway colonists will need human expertise, various products and perhaps also raw material from the inner regions of the planetary system. The Kuiper and Oort pioneers will have to go it alone, however. Perhaps they will engage in competitions for distance and velocity; how far and how fast can we go?

Human presence in space will expand towards the stars, but it will be neumannbots and not humans of flesh and blood who will actually reach the further limits. As opposed to the established order in science fiction novels, it will take thousands — if not millions of years before the inhabitable planets of neighboring stars are reached. The Oort Cloud dwellers will help by building extremely strong telescopes, providing the rest of us with useful information that will complement the knowledge we already have. For a three-dimensional image of space, telescopes positioned at great distances are incredibly important; with two eyes it is possible to see depth and the further these eyes are apart, the better their sense of depth will be.

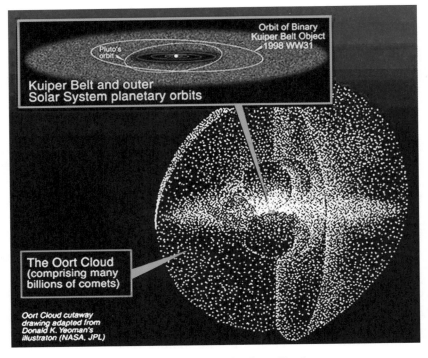

The Kuiper Belt and the Oort Cloud.

As an aside, I believe we could let our imagination run wild with ideas of what we could do with the numerous asteroids. Not all the worlds we discover will have to be colonized; there are lots of fun things we could do with them. In a micro gravity environment, the wildest constructions are possible without collapsing under their own weight. For example, entrepreneurs could decide to turn some of the asteroids into enormous entertainment parks, filled with rides that would be utterly unthinkable elsewhere. Or artists of various vocations could make something beautiful out of an asteroid, either by physically sculpting it themselves or reworking it remotely using neumannbots. One might transform an asteroid into an aesthetically pleasing but incredibly complicated mathematical figure such as a *fractal*, a mathematical pattern containing millions of repetitions of itself at ever tinier scales. Others might create wild and bizarre sculptures, while artists from Japan, Korea or China might be inspired to convert their asteroids into gigantic Buddhas — entertainment, art and religion will always continue to be important human drives.

There will undoubtedly be scientists who will want to build special scientific devices on some asteroids: gigantic telescopes dwarfing all those we already have, which will help us discover new icy worlds in the Oort Cloud, or assist us in our research of planets in other star systems. But they will also help satisfy our curiosity towards all those wonderful worlds located at even more mind-boggling distances in our universe.

Of course, developments could take quite a different turn. I mentioned all those unbridled possibilities regarding communication and computer techniques. Would people really want to make those dangerous space trips themselves? Perhaps virtual reality will be perfected to such an extent that we could take a stroll amongst the stars right from our living room. Enormous data banks could digitally store all the newly discovered sceneries: we would climb into our virtual spaceship and make whatever voyage we wish, easily reaching velocities thousands of times that of light. Nothing will be more comfortable. The adventure of space travel will mostly have been lost, but perhaps the virtual trips will be so realistic that the average easy chair astronaut won't even notice that what he or she is doing is not the real thing!

chapter 14

The Genes

IN THE PREVIOUS chapter, I showed how to reach for the stars without breaking any natural laws, but there are just as many important issues to address closer to Earth. I already mentioned the information revolution and nanotechnology, which will lead to as many unpredictable developments.

Traveling to distant stars and planets appeals to everyone's imagination, but the voyage to the small is just as spectacular. Neumannbot equivalents in the world of the small already exist: these are bacteria and algae (viruses are even smaller, but they can only exist in other living creatures, while bacteria and algae can exist independently). What we don't like about bacteria is that they are too independent: they don't take any notice of what we humans want and we often have difficulty controlling them.

You probably think that the nanotechnology developed by information specialists, on the one hand, and bacteriology, on the other, are completely different areas of expertise, but that's not quite true. They both focus on the world of the tiny, the nano scale, and they are becoming more and more interrelated. In the longer term, perhaps even a fusion is feasible.

For researchers in this area — the grey area between nano-technology and biology that does not even have a name yet —

a lot remains to be done, but a start has been made. The first signs are positive. I am referring to modern techniques, among other things, used by researchers to establish the order of amino acid molecules in DNA chains of living organisms. Organisms' DNA chains contain practical information regarding hereditary characteristics.

In these stretched out and entwined molecules, nature has preserved entire programs that determine the make-up of all organs in the greatest detail. The amino acid components (of which there are four, usually abbreviated by the letters A, C, T and G) are used by nature to create lengthy pieces of prose in our DNA, in total a text of about three billion letters, equivalent to about one gigabyte. It was an enormous challenge for biologists and chemists alike to decode the DNA of highly developed organisms, particularly that of humans. But they triumphed.

In the year 2000, the first rough draft representing the human DNA sequence had been created, the so-called human genome, but only in 2003 did biologists dare to trust the composition completely. Remarkably, our code is no longer or more complicated than that of a chicken or an earthworm. Moreover, the codes are 90% identical! From person to person, irrespective of race, the differences are less than 0.1%.

This result immediately triggers lots of questions: how does nature read the code and what do all those letters mean? Are they arranged in a specific way and would we be able to lay bare all the details? Where is it written that we have five fingers on our hands and five toes on our feet? And where can we find what they look like? Our body is one large information storage and processing system. All sorts of hormones, added to our body by our genes, pass along signals. They react to stimuli from the environment, they ensure that the muscles we use regularly grow stronger, and so forth. How all this is done is largely unknown. DNA appears to resemble a huge computer program that is filled with all kinds of complicated subroutines.

The steps that are still required — and I don't see why these couldn't be taken in the not-so-distant future — include the following:

- We should be able to read DNA codes much faster, until it resembles reading a computer disk. Nanotechnology will play a large part in this.
- We have a lot to read: we want to know the genomes of all living organisms on Earth, and I would recommend starting with the genomes of all those organisms that are threatened with extinction.

Only in this way will we learn to understand how genes work. I think the need will arise to identify the genome of every human being, which would considerably improve making diagnoses and prognoses, and determining treatments.

We would then also be able to grow a 'tissue reserve', which would eliminate the dangers of rejection; a body would, after all, recognize its own tissue. Now I am well aware of all sorts of issues that will be raised by this technology. How are we going to protect an individual's privacy? I won't say much more on this point, but it should be evident that the privacy of the individual should be well protected, and I assume solutions for this will be found, somehow.

But there is more. After learning to read, we must learn to write. We can then experiment to our hearts' content. This will have even greater ethical implications, which we will have to face.

The movie *Jurassic Park* assumes that the DNA of dinosaurs is still preserved, intact, in mosquitoes found in amber. In reality, these mosquitoes would have been in an advanced state of decay, and what remains of the blood DNA that was still present in their bellies would not have been viable. But it might have been that traces of DNA were still there, and that advanced detection techniques could have made something out of it. This, combined with a huge amount of knowledge of how DNA works, might give researchers the opportunity to retrieve the information stored in this highly defective DNA. In other words, it is not totally unthinkable that animals that have long since been extinct might be resurrected, including dinosaurs.

The rest of the movie is not really worth our time. I don't know whether there will ever be any such desire, but even if we wanted to resurrect pre-historic animals, we could do so in a zoo. They would hardly pose any real danger to humans or contemporary animals; on the contrary, it would be difficult to protect the pre-historic beasts against today's aggressive life forms, which, after all, have evolved much further. As an aside, there have already been thoughts to bring back the dodo and the woolly mammoth, so it is not as if we are that far off such developments.

It would be much more fascinating, though, if species could be created that never existed and which would never have emerged naturally through the evolution process. I am talking about animals, plants and single cell organisms that could assist humanity in all sorts of ways. Think of the food industry: it is really barbaric the way in which we stuff all sorts of soft and furry animals into too small cages, force feed them and then convert them into food for ourselves. Instead, wouldn't we be able to create slabs of meat without brains, which therefore wouldn't feel a thing? To produce milk without a cow? Eggs without the intervention of a chicken? Fruit and vegetables could be grown quickly and efficiently, and adorned with whatever flavor we want.

Of course, fruit and vegetable growers have been working on this for a long time, but they can only use existing plants and animals, and through cross-fertilization optimize certain hereditary traits. Sometimes scientists manage to transfer the genes of one organism into those of another, mainly through trial and error. If we knew how to create totally new hereditary traits, in exactly the right way, the fruit and vegetable industry would be transformed beyond recognition.

The food industry is not the only sector in which we could use artificial plants. I was talking earlier about filtering saline water to get fresh water; perhaps in good time we will be able to fabricate genes enabling us to create a new plant that, with the aid of sunlight, is able to transform saline water into fresh water. This would help us prevent further desertification in the future and perhaps even influence our climate, as described in Chapter 8.

Talking about solar energy, we could suggest something similar: design a plant that can transform sunlight directly into fuel for cars, or other plants that manage to produce electricity directly. Granted, this is a bit far-fetched, but genes in organisms that can produce electricity already exist, just look at the electric eel. So once we understand how genes work and how they are positioned correctly, then this might not be so impossible after all.

Other practical implications would be in our continued biological battle against plagues and pests, such as mosquitoes. Mosquitoes should not be terminated altogether because they have important functions in our environment, but something could be invented to dissuade them from preying on humans.

A long time ago I read a science fiction story in which genetically modified house pets were intelligent enough to talk to humans. In itself, the idea is not that unbelievable, but to be able to speak, far-reaching modifications will have to be made to the jaw and the larynx, which would result in a cat that will no longer look like a cat. Moderate transformations to increase the intelligence of pets is thinkable, but I have already told you how difficult it is to produce intelligent computers, so the foreseeable difficulties with trying to pinpoint human intelligence and to determine how we could alter this genetically will, in all likelihood, discourage us from carrying out these sorts of procedures for a little while longer. Of course, one could wonder whether we really want intelligent pets — ok, I suppose it would be nice if we understood them a bit better and they us.

An alluring idea for science fiction is a firm specializing in farming butterflies and flowers which, for a huge fee, could emboss a company logo on a butterfly's wings or a tulip's petals. The implication that people won't know where to stop with this sort of folly, is not at all too far-fetched.

Letting our imagination run away with us even further, we might expect that genetic engineering will become an integral part of space colonization. People who plan to live on the Moon and have offspring there will want to modify their genes in such a way that they feel most comfortable at one-sixth of the Earth's gravity. Colonists on asteroids or other space habitats will need to genetically modify their own genes as well as those of their livestock even further.

Or how about growing our own houses? One science fiction story I read told of mansions where if you wanted to hang up a painting, a little bit of DNA was injected into the walls, so that it would grow a little hook.

As this author is a physicist and not a biologist, he is not constrained by as much knowledge in this area as his colleagues in the field of biology would be, who would surely counter this idea for the future with lots of sensible arguments. It is not as if our genes are architectural drawings in which we can easily scribble some amendments. The way in which an organism is programmed is immensely complicated, and a logo on wings will certainly be no mean feat.

And even if a plant or animal is crafted, it will have to defend itself against all sorts of parasites and diseases, and it must be able to reproduce. In addition, we will have to be very strict to ensure that these fabricated plants and animals will not endanger our existing flora and fauna. This will prove quite a hefty bone, but I have high expectations of human ingenuity and I am convinced that there are a lot more applications to be discovered for biotechnology or gene-technology than most people realize. The sky is the limit!

As I said, we expect a lot from nanotechnology, which we will use to read and write DNA. But this 'cross-fertilization' might work two ways: nanotechnology will also be able to use the incredible variation of a DNA molecule. This is already happening.

Microscopic engines and detectors can be made out of DNA, and perhaps DNA itself could be used as computer memory storage; after all, DNA is nature's way of transmitting information at molecular level, in a far more sophisticated manner than today's computers.

chapter 15

Pulling Hard

IN THE MEANTIME, another silent revolution is taking place: in the area of materials. Materials are becoming better: lighter and stronger. This is evident from the construction of new buildings and bridges. One of the most important components of these new materials is fiber.

The tensile strength of fibers in relation to their weight, that's what counts. The potential of a new material is in direct correlation with the strength of the fibers that have been used. An Internet search uncovers something remarkable; the strength of a fiber is indicated in terms of grams per denier. What is a denier? A denier, as it turns out, is the weight in grams of a string of silk with a length of 9,000 meters. A fiber with a strength of one gram per denier means that the fiber is strong enough to carry its own weight, if it is not longer than 9,000 meters. The Internet mentions very strong fibers of, for example, eight grams per denier, and on one site a Kevlar fiber is mentioned with a strength of 23 grams per denier. Such a fiber could carry its own weight even if it is 207 kilometers long!

These numbers indicate how long the greatest span of a bridge could be, if it were completely built of these fibers, or how high a kite could fly if the string were made of this fiber. Of course, a bridge also has to carry the weight of the road surface and cars, but a span of dozens of kilometers is definitely possible. One of the strongest natural fibers, by the way, is spider's web, or gossamer, which is almost as strong as Kevlar. We have not yet identified a method to artificially fabricate gossamer, but you understand how I would solve this: we replace one of the genes of a silk caterpillar with that of a spider, and have caterpillars weave gossamer. Ordinary silk is much weaker.

The strength of fiber is thus easily expressed in terms of how many kilometers of its own weight it is able to carry. The thickness of a string is immaterial, a thicker cable may be stronger, but will also be heavier. In practice, the maximum length of the strongest fiber is about 200 kilometers. Compare this with the strongest brand of steel: 13 kilometers.

I should remind you that the cylindrical space habitats of Gerry O'Neil had a maximum diameter of 19 kilometers. That is the maximum size that can be made to hang together when normal steel is used, assuming that we wish to generate an Earth-like gravity on the inside. At every point on the cylinder, the two halves support each other, so that boils down to each portion carrying about 9.5 kilometers of its own weight. The surplus of strength goes into holding the weight of whatever is in the cylinder, plus the air pressure.

But what would the theoretical limit be for fiber and cables? A lot more than the above. Diamond is so strong that it could carry its own weight at a length of 3,800 kilometers. But the strongest material would be produced by rolling up carbon atoms in a fishnet construction. These so-called 'nano tubes' are currently being researched. They have many special characteristics, such as being superconductors at low temperatures. Their most remarkable property is, however, their strength: they can carry their own weight in cables made of their own material at a length of 11,000 kilometers. That is more than the radius of the Earth. This material could be used to create a cable that connects the Earth with interplanetary space. More precisely, a satellite that orbits the Earth in exactly one day, such that it is in what is called a 'geo-stationary' orbit, could be connected with the Earth's surface with such a cable.

As an aside, I am using the most optimistic figures here. Other numbers are circulating in the technical literature as well as on the Internet, suggesting that these materials would be a lot less strong than I am putting forward here. My numbers would be valid in the most optimistic scenario.

I remember an event, where I spoke to a specialist in the area of materials about fibers, and he quickly understood where I was heading with my questions. Bored, he looked at me; *"there's another one fantasizing about a space elevator,"* I could hear him thinking. *"You must realize that there is an enormous difference between theory and practice,"* he said. And indeed, in reality we have to consider material defects.

The occurrence of defects is unavoidable. The reason why materials are so much weaker in practice than their theoretical strength might suggest, is because ruptures appear. If ONE atom is out of joint, an avalanche starts, and all the neighboring atoms let go as well. The material ruptures and the cable breaks. Our goal should always be to keep those ruptures to a minimum by building barriers, but even the smallest tear weakens the material considerably.

Ok, that's the practical problem: not one atom must be out of place and if that does happen, then the consequences must be minimized. This, it appears to me, is a task for the nano technicians. At this moment in time, it is possible to construct nano tubes, but they are small and short. We will have to devise a technique to weave them, just like rope is weaved out of cotton. Even if only half its maximum strength is realized, say 5,000 kilometers of its own weight, then this would help to materialize all kinds of revolutionary applications, as well as a space elevator. Of course, I imagine that cables of such an extraordinary strength will be put to good use in space. A cable that could tug its own weight could be tied between a fixed place on Earth, such as along the equator, at one end and a geostationary satellite at the other.

Geostationary satellites are normally located at a distance of 35,783 kilometers from the Earth's surface. This is the distance at which the Earth's gravity is exactly strong enough to pull a satellite along, synchronic with the Earth's rotation. If we want that satellite to carry a cable that holds an elevator, the distance will have to be a bit longer; at least 40,000 kilometers. The Earth will tug hardest at the bottom part of the cable because the rest is too far away. If you make some further calculations, it becomes clear that a cable that is able to carry some 5,000 kilometers of its own weight would suffice. However, the thickness of the cable has to be adjusted very precisely to what will actually be needed at every point along its length: a bit thicker at 35,000 kilometers high and quite a bit thinner right above the ground because, close to the Earth, the cable will only have to be able to carry the weight of the elevator, not the rest of the cable.

A space elevator is an alternative method of reaching interplanetary space, without the fireworks of a huge rocket. This is what I was getting at in Chapter 1; we attach a cable from a geostationary satellite to Earth, and along this cable we move an elevator up and down. Now remember my remark in Chapter 2, that lifting any kind of payload all the way up to interplanetary space costs a lot of energy no matter what technique you use. Here, however, there is a possibility to save a lot of energy, if you manage somehow to connect the ascending elevator to the one coming down.

This idea of the space elevator is not entirely new. Already in 1895, Russian scientist Konstantin Tsiolkovsky imagined building a structure *'much higher than the Eiffel Tower'* and connecting it at the top to a cable that was attached to a *'celestial castle'*; that must have been what we would now call a geostationary satellite. Spaceships would be able to pull themselves up into interplanetary space. Particularly well known is the meticulous description that science fiction writer Arthur C. Clarke provided in his novel 'Fountains of Paradise' in 1978. On a mountain near the equator, a miles-high tower would be built and connected to a cable that would be as strong as steel but as thin as dental floss. Now we know that this is possible, at least in theory.

The American space agency, NASA, also realized its potential and assembled a group of scientists in 1998, who were tasked to do research on this project. It was concluded that a base tower of at least 30 kilometers in height would be required. We know that, in principle, this is possible with our current building materials. There, high up in the stratosphere, the cable would ascend to the geostationary satellite. The space elevator must not touch the cable, but would be able to pull itself up with the use of magnetism, just like a suspension train that floats on a magnetic field. The researchers opined that this must be possible within the next 50 years or so. The geostationary satellite would have to carry a heavy counterweight to be able to supply the necessary force. A small asteroid might be used, it was thought.

A problem with carrying out this project is the large amount of 'space garbage' orbiting the Earth, originating from the numerous artificial satellites that have been launched over the years and the various parts that have come loose. Many satellites have become defunct entirely, thus adding to the junk in space. One idea under consideration is to hang the cable loosely, so that it can avoid orbiting garbage. I think it might be better to clear away all the debris first; this will have to be done sooner or later anyway. All the rubbish, whether nuts and bolts, slivers or entire artificial satellites, will have to be detected and caught, one by one. I imagine that in the era of superior information technology that's around the corner, these sorts of chores will be child's play.

It is clear that we are actually still far removed from making this space elevator, with the most important stumbling block being the need for strong cables dozens of times stronger than anything we have at present. One could wonder whether we should start with a slightly less ambitious plan.

Well, the Moon also has a gravitational field and you know by now how important I rate the Moon in terms of future space adventures of any real consequence. The Moon's gravity is much lighter than that of Earth and a cable that could carry 200 kilometers of its own weight on Earth would already be enough to connect the Moon with interplanetary space. An object would then be in orbit around the Moon. We need to put a satellite or space platform in the so-called 'Lagrange point'. These are points in space where a vehicle will remain in a set position in relation to the Moon or the Earth. The closest point is more than 60,000 kilometers away from the Moon's surface, but while the cable would have to be longer than the one holding up an elevator from Earth, less would be required in terms of strength.

Apart from our elevator, cables and fibers would have all kinds of interesting applications in space — connecting one spacecraft to another, for example, or attaching a spaceship to an asteroid.

This brings me to artificial gravity. You know that in television images, space travelers always float freely in their spacecraft, or at least as long as the engine is turned off. As the engine uses a lot of energy and fuel, it will mostly be off. Of course, weightlessness is not a natural environment for a human being and this could cause problems, such as bone decalcification. Our bones and muscles have very little to do in zero gravity, and become seriously weakened.

However, it is possible to produce artificial gravity. In science fiction, this is realized by having large spacecraft turn around their axes, just like O'Neill's cylinders. The occupants would be pushed outwards to the sides of the cylinder, which then seems like it is the floor. But it would also be possible to connect two spacecraft. With a cable of 500 meters, we could let the spacecraft circle around one another with a speed of two revolutions a minute. In both spacecraft, the far ends will serve as floors.

Such a mechanism would already be applicable to the international space station (ISS), which is currently orbiting the Earth. At some distance from the centre, two modules could have been attached with a cable, one on each side. If the cables were each 250 meters long and had a rotational velocity of two revolutions a minute, then the gravity in those modules would be similar to that of the Earth's surface, very comfortable for the crew who live in ISS for a long period of time.

This kind of construction is definitely an option for the space capsules on our journeys to Mars, so it won't be necessary for future astronauts to make the trip in zero gravity. That is not unimportant, because the trip will take approximately eight months. The rotational motion would be so slow that people would quickly get used to it, even though it might be very noticeable and perhaps irritating at first.

Working with kilometers-long cables in space is a tricky undertaking and several attempts have already failed. A test with a satellite linked to a space shuttle was unsuccessful because electrical currents ran through the cables. That would not necessarily have been a problem if there hadn't been a weak spot in the cable, which had a greater resistance than all other sections of the line. The temperature rose at this spot, weakening and eventually breaking the cable. It is clear that the use of cables in space has its safety concerns and we would have to ensure that the cables couldn't break.

When the era of colonization of asteroids and small planets starts, whether by humans or robots, cables will play an important part. Two objects can be linked with cables and brought into motion, after which the objects can be slung in the desired direction, by breaking the connection between the two at exactly the right moment. Even asteroids can be maneuvered this way. The asteroids' large mass means that the change in velocity might only be small, such as one centimeter per hour, but even that speed can be very important. Let's say, for instance, that there is an asteroid that is posing a danger to Earth or one of our space colonies. Then a small change in its pace, as long as it was started long enough in advance, would suffice to turn the tide. I will return to this topic in Chapter 17.

If you were planning to let your own fantasy run wild regarding all the applications of cables in space, I have to draw your attention to one serious constraint, which can be perceived as a law of Nature. I have previously indicated the quality of the cable in terms of the number of kilometers of its own weight that it is able to carry in the Earth's gravity. In space, there is no gravity. Here, we indicate the quality of the cable in km^2/sec^2, or velocity squared. This figure is the outcome of the number of kilometers of its own weight times the acceleration of whatever planet's gravity. Earth's gravity has a strength of 9.8 m/sec^2 (the 'acceleration of Earth's gravity' refers to the fact that when an object is dropped, it accelerates, dropping faster and faster until it reaches almost 10 m/sec in one second). Nano tubes, one then finds, have a quality of about 100 km^2/sec^2. That roughly equates to the square of the Earth's escape velocity, and means that they would be suitable for building a space elevator. With Kevlar, the quality lies at around 1 km^2/sec^2, while steel is at around the square of 350 m/sec, or more than 1,000 km/hour. The contention is that, if two objects are connected with a cable, the relative velocity of both ends may never be much larger than the speed that is characteristic for that particular cable strength. For example, the ends of a nano tube must never rotate with a larger velocity than 10 km/sec, as explained above. This also determines the speed limit of a car or train on wheels fortified with cables; exceeding this speed limit may cause the wheels to explode.

The same logic applies to flywheels. Flywheels are used to store kinetic energy, for instance in vehicles. When the driver uses his brakes to reduce speed, the energy can be used to accelerate the flywheel, which in turn can redeliver that energy at a later time when the driver wishes to regain his velocity. In principle this procedure would require no more fuel in total than when the velocity had been kept constant — a big gain.

If the flywheel is made of steel, as explained above, it may never rotate faster than a few hundred meters per second. That limits the amount of energy that can be stored. This is why flywheels made of steel cannot store as much energy as batteries, such as those we use in our cars, let alone gasoline: chemical reactions can make molecules move at several kilometers per second, and it is the square of this velocity that counts as energy. So yes, chemical reactions should be able to store much more energy than a flywheel.

Contrary to what one might have thought, we know that the storage capacity of a flywheel per unit of weight is independent of size. Nevertheless, if size and weight do not matter, a flywheel could have definite advantages. This is because the efficiency of the conversion of energy to and from a flywheel can be very high.

Returning to the subject of space travel, we find that asteroids and spacecraft can only be linked with cables if the velocity differences remain below the limit imposed by the quality of the cables. Orbital velocities are often dozens of kilometers per second, so two objects can only be linked by a cable if they are in nearby orbits, where their natural velocities would be almost the same.

chapter 16

Aliens

LET ME ALTERNATE fiction with a true story. Sidney Coleman was a well-known and influential scientist in my area of expertise, affiliated with the reputable Harvard University. Sidney happened to be a great fan of science fiction. A good friend of his was Carl Sagan, the author of widely-read popular science books and television series, such as *Cosmos, a Personal Voyage*. Sagan also set up a system with which he hoped to discover extraterrestrial civilizations, by meticulously analyzing radio signals emanating from the cosmos. This was the so-called SETI-Project, *Search for Extra Terrestrial Intelligence*.

In the 1960s Coleman and Sagan regularly got together to chat about science fiction. One day they found themselves in a nice gentleman's club. "*Sidney,*" Carl said, "*I have an interesting question for you. Soon, the first astronauts will land on a celestial body, the Moon. No one knows whether life has ever existed on the Moon and whether there might be micro-organisms there, dangerous to our health. What if these astronauts unknowingly bring these organisms back to Earth? What kind of precautions should NASA take to protect us against this potential hazard?*" This problem was further deliberated over a good glass of wine. Various ideas were put to the table. Of course, neither had the faintest idea what these micro-organisms would look like, and logically and scientifically, such organisms couldn't really exist. But what if?

It was a thought-provoking discussion. Vaguely they agreed that returning astronauts should perhaps be kept in quarantine for a few days, just to be safe. For trips to the Moon this would merely be a formality, but once we start to travel to Mars, we have to be more careful to avoid contamination. Particularly Earthly contamination of Mars, but also the other way around. A protocol should be developed to which everyone must adhere. "*Let's insist that the chance for microbial contamination from one celestial body to the next should be less than 0.1%*," was their temporary conclusion.

Sidney regarded the entire conversation as yet another reverie about science fiction. Imagine his surprise when a draft manuscript fell on his doormat a month later. 'Spacecraft Sterilization Standards and Contamination of Mars', by C. Sagan and S. Coleman, to be sent to the *Journal of Astronautics and Aeronautics*. The manuscript argued that the norm for the possibility of cross contamination from one celestial body to another should be less than 0.1%. Further, it was set out how astronauts should be put in quarantine and, lastly, how the chances of survival for every individual organism of a life form from one planet to the next should be less than one in 10,000.

The article was published and because it was the only scientific piece on this issue, COSPAR (Committee on Space Research), the special commission from NASA that guards over safety aspects of space missions, adopted the standards. A special purpose building was constructed, the Lunar Receiving Laboratory (LRL) in Houston, Texas, where astronauts returning to Earth were quarantined, while any rocks they found were examined separately in a sealed room. Of course, nothing of any consequence was ever found. And thus, how to protect humanity's fate in the face of a potential worldwide disaster was decided upon, while chatting over a good bottle of wine. End of story.

The probability that a life form of whatever nature has come about on any planet or moon other than Earth is tremendously small, in my opinion — even on Mars and Jupiter's moon Europa, which have been the subject of loads of speculation. It is imaginable, though, that primitive life forms have developed somewhere many light years away from us. If any intelligent life has come from these primitive life forms, it will in all likelihood have happened at great distances from us. I will tell you why I think so.

I have already explained what living organisms really are: they are neumannbots that spontaneously came into existence through natural processes. It all started with bacteria and even more primitive life forms, which after billions of years evolved into the unbelievably complex wealth of life we now know on Earth. I have also explained how extremely difficult it is to design and build neumannbots. It is my belief that this could only have happened spontaneously on Earth as a result of the unique combination of virtually ideal circumstances: exactly the right temperature, raw materials and environment, as well as a perfect combination of various challenges and cosmic events that have shaped our evolution.

For every given planet, the probability of exactly such a thing happening or having happened is extremely small. Events that have occurred on Earth and which have led to the spontaneous development of life could occur on other planets, but the required set of circumstances are extremely rare. That's why I believe that there are very few planets in neighboring parts of our universe where such a miracle has taken place.

There is a fact of natural law that is often sidelined by science fiction authors; it doesn't only apply to us, but also to alien visitors — those little green men who come to visit us in their spaceships, but who, apparently, hide themselves from all Earthly scientific researchers. The reason why these alien visitors continue to escape our detection is crystal clear to me: they don't exist. Or at least, not here.

I have already made a wager with a colleague, insisting that no life form whatsoever will be found on Mars and I am prepared to place the same bet for all other planets and moons in our solar system. My guess is that the closest planets where living organisms have sprung up are hundreds of light years removed from us, and that intelligent beings must be sought much, much further out than that. At such distances, there are so many stars and planetary systems that the odds in favor of something extraordinary happening there may be slightly higher.

This statement, I admit, is not one based on hard science, but rather one of personal surmise. But even if I am off by a factor of 100 — and that is easily possible — then aliens will still have to travel dozens of light years to reach us. Even they won't be able to travel faster than thousands of kilometers per second and they won't, just like us, want to make such travels themselves; they will construct neumannbots and send those into space. Therefore, if we ever observe any aliens, they are likely to be robots, an intelligent manufactured life form entirely adapted to make long-distance space voyages.

Imagine an alien life form that found the power, endurance and motivation to embark on voyages that take many thousands or even millions of our human years. They would have to protect themselves against impacts of meteorites, against radiation damage and other sources of aging that would accumulate to staggering intensities over such time stretches. Clearly, spacecraft making such a trip must contain all instruments to repair whatever damage it may endure over time. In my estimate, such ships must contain entire communities of specialized robots. The hardest hurdle to me seems to be motivation: why try to do this?

Incidentally, this is also an argument against the so-called 'panspermia' theory, the idea that primitive beings may be roaming around in outer space, contaminating all planets so that life can develop there. Radiation from high energy particles and from ultraviolet light is far too hazardous to allow DNA or similar molecules to retain and spread the information they contain, fighting against the odds of having to spend millions of years in space before reaching a susceptible planet. The conditions in outer space cannot possibly be compared to the protective properties of a well-chosen terrestrial atmosphere. Evolution cannot take place freely in space or in those small celestial bodies (comets, asteroids and such) that are sufficiently numerous between the stars, to be held responsible for an efficient dispersion mechanism from one star system to another. Scientifically speaking, such a theory is not impossible, but to my taste extremely unlikely to hold true.

Yet there is a way that life can spread, and that is by being intelligent. As I argued in Chapter 13, I believe that neumannbots will slowly but surely conquer the space around us. They will be able to travel far beyond Pluto, jumping from asteroid to asteroid, from the Kuiper Belt to the Oort Cloud and eventually to other stars from there. As our neumannbots won't be able to travel much faster than dozens of kilometers per second, this will be the speed with which they will eventually gobble up the entire Milky Way. This means that within a few million years, a considerable part of our galaxy will be occupied by neumannbots.

This calculation might not be completely off the wall. Paleontologists have researched how fast the human civilization spread over the Earth. It is apparent that new inhabitants with innovative techniques usually spread with a speed of dozens of kilometers per generation. That was simply the speed by which these inhabitants colonized their surroundings; offspring settled a couple of kilometers away from their parents. The same thing will be true for neumannbots; they will spread with the same speed as they set up colonies, which is at the speed at which one neumannbot manages to reach the next inhabitable asteroid — with a speed of about a few kilometers per second, or a light year in 100,000 years.

Neumannbots that aren't interested in traveling from star to star all at once, but instead hop from asteroid to asteroid or from comet to star, should nevertheless be able to spread over the entire Milky Way in less than a hundred million years or so. Maybe I am off by a factor of ten, but even if they take a billion years, that is still very fast. The Milky Way is almost as old as our Universe, more than 13 billion years.

If there had been other planets in our galaxy where human-like civilization had been founded, then you'd expect them to have similar ideas. Not everything takes 13 billion years to develop. But if there exists a planet where more than a billion years ago an intelligent life form sprung up similar to ours, then their neumannbots would have reached us by now. Despite the countless reports of UFOs, I don't think this has actually happened; with due respect for the UFO watchers, I cannot imagine any good reason why such neumannbots would not have left their trace, which would have been obvious for all to see, anywhere in our solar system. They have not. Therefore, I must conclude that it is unlikely that intelligent life forms have settled anywhere in our galaxy. On the other hand, there are many billions of galaxies in the visible part of our Universe. There may well be civilizations in many of them. Due to the millions of years that signals take to reach those galaxies, a two-way communication channel with them is out of the question, but detecting such a civilization — who knows — might be possible.

The arguments here were first raised long before I did, by physicist Enrico Fermi. *"Where is everybody?"*, was his rhetorical question regarding aliens. I completely agree with him. It is, of course, imaginable that it is fundamentally impossible for any type of neumannbot to cross interstellar distances, since natural laws are not particularly permissive when it comes to hopping from comet to comet. Naturally, it could be that other human-like civilizations have decided to abandon the idea of building neumannbots. But why? I don't understand why that would be.

chapter 17

Playing with Planets

UNTIL NOW, I have limited myself to only those adventures that humanity could plunge into without blatantly offending too many natural laws and without making too many unreasonable investments. Everything is in accordance with what we now know, or rather, what *I* now know, of nature's laws.

Let me take this one step further. Let's assume that humanity has occupied all planets and moons with neumanbots or cambots, and that adventurous colonists have inhabited all worlds in our planetary system, where some residents have genetically manipulated themselves so as to better adjust to their new habitats. After all, many moons and planets are blessed with only a fraction of the Earth's gravity, and the colonists' bones should not suffer too much from this. The colonists must have added something to their genetic make-up to resist bone calcification. Further, they may have grown new types of crops and farm animals, properly adjusted to whatever new day-and-night rhythm they may have to cope with and so forth. Of course, Venus would be beyond our pioneering spirit because it is too warm and the planet's atmosphere is too dense and toxic; because of gigantic amounts of carbon dioxide, the greenhouse effect there has run wild. And then there is Mars, but the Red Planet is too cold and its atmosphere too thin; Mars does not have enough greenhouse gases.

But imagine we have now advanced a million years. Perhaps terraforming has been applied, perhaps not. Is there nothing else that can be done? Could we not move Venus a little bit, and Mars too, to improve their habitats? These questions concern a development that is the direct opposite of our accomplishments in nano technology; we don't just want to manipulate incredibly small objects, but also amazingly large ones. In the world of small planets, asteroids, there is microgravity. That's why we will be able to use machines there that are much larger than the diggers and blast furnaces on Earth. Why should we not conquer the world of the large just as well as the world of the small? Asteroids may be a good place to start; they come in all sizes.

Let me answer the question straight away as to why this will be difficult; it's because of time. Generally, large objects move much slower than small ones and therefore all changes will, in Earthly terms, be very slow. The enormous amount of steel that we plan to harvest from asteroids to build Gerry O'Neill's cylinder-shaped space colonies, for instance, will have to be processed in gigantic blast furnaces in space. If they work at all, they will work only very slowly compared with terrestrial furnaces, just because of their size.

But anyway. How could we influence the orbits of planets? Let me first describe what I have long assumed to be a possibility in principle, until I stumbled upon scientific work that put it all in a totally different light. Starting with the smallest asteroids, we connect them with cables to smaller rocks that also roam around in space. These rocks are small enough to be moved around using rockets. Asteroid and rock together form a sling, and by severing the link between them at exactly the right moment, we can create small but significant changes to the asteroid's orbit. Just enough, for example, to engineer a crash between two asteroids or, even better, to guide them towards the gravitational field of a larger planet. If another change in velocity is created, at exactly the moment when the asteroid approaches a larger planet, then the effect will be greatly enhanced by the gravity of that planet. In this way, we can create significant changes to the orbits of asteroids.

We will now ensure that our asteroid comes near a larger planet, such as Venus or Mars. Planet and asteroid also form a sling, but one of much greater magnitude; because each time an asteroid approaches a planet, they exchange some of their kinetic energy. More precisely, there are two physical properties we want them to exchange: kinetic energy and the orbital angular momentum. The latter refers to the rotational motion of the planet around the central point of its orbit. If you are not familiar with these principles of mechanics, I won't bore you with details of the relevant formulas. The main point of this idea was that it could be a way to move asteroids around, and they in turn could be used to influence the motion of planets, away from their original orbits.

We would set up teams to drive their own asteroids. All teams attempt to transfer as much kinetic energy and orbital angular momentum as they can from one planet to another. We need at least three planets for this, for example Venus, Mars and Jupiter. Jupiter is too massive to move from its orbit, but its kinetic energy and orbital angular momentum are useful to us.

My idea, which could take millions of years, is to move Venus a bit further away from the Sun, and Mars a bit towards the Sun through this method. Then both Venus and Mars might become inhabitable, or at least a bit more hospitable, for us. Once in their new orbit, these planets could begin writing their own fresh history of evolution, perhaps even creating life, or at least being able to sustain it if it is introduced by mankind.

In order to move planets into new orbits, it is crucial to measure and calculate the exact orbital characteristics of both planets and asteroids with enormous precision, but that should not be a problem: the distance between the Earth and the Moon has presently been established with a margin of error of only four millimeters. That will become microns in due time and in the somewhat more distant future, nanometers.

Now this was the original idea. Reality is, of course, quite a bit more complicated, because all planets and moons influence each other's movements. These influences are minimal, but in a time span of thousands of years they are noticeable: that's why the planet Mars was closer to Earth recently than it has been in over tens of thousands of years. Mars's orbit is not constant, but is adopting an increasingly elliptical shape. This is caused by the current positions of the planets. In 10,000 years, this eccentricity will disappear again. Consequently, the various planets in our solar system do show variations in their orbits, but return to their original positions in due time. In the distant past, it took millions of years for the system to stabilize, and what this means is that, whatever we try to do, the present situation will return after a few thousand years. So, our laborious attempts to move Venus away from the Sun, and Mars a bit closer, will fail unless additional measures are taken.

To be precise, within the system, one encounters a certain level of *chaos*. The way in which orbital parameters fluctuate is not entirely predictable; it is much too complicated for that. In fact, it resembles the behavior of the weather on Earth; over long periods of time, the *average* weather is quite accurately predictable, but in the short-term the weather fluctuates drastically. It is unpredictable because every variation, however small, will grow into a complete change of the weather pattern. Every rounding error we make in our calculations, no matter how tiny, will result in our predictions being wrong within a certain period of time. This is what we call 'chaos' in science.

The planet whose orbital elements fluctuate most wildly is Pluto. Pluto is likely to remain in its orbit on average, but it cannot be completely excluded that it may move towards a totally different orbit or even leave our solar system entirely. And it could all depend on that same little butterfly that spreads its wings and noticeably changes our weather a few weeks later.

Calculations have been carried out to identify what would happen to our planetary system if a planet the size of Mars or the like was added. To the surprise of the researchers, it appeared that the alien planet would in due time be expelled from the group by the other planets, which would return to their original setup.

Stability has its conditions: small changes are annihilated, but if the orbits are changed too drastically, stability may be lost. Planets could crash or become moons of one another. If we were to carry out my idea, the stability of our planetary system might be disrupted. Even if we were to leave the Earth out of it, our orbit could change and a crash between the Earth and Mars or Venus would be difficult to avoid.

Not only should we avoid crashes at all times, but planets should also never get too close. If Venus or Mars came as close to the Earth as the Moon, the forces of the tides would cause devastating floods and earthquakes. Orbital elements must not exceed certain boundaries because, if they do, we lose all control over them.

Thus, chaos continues to be a problem, but it might also be its own solution. An extremely advanced civilization might be able to use this instability to ensure that planets are rearranged into acceptable orbits by causing highly complex pre-calculated interruptions to the system. We assume here that scientists are able to measure and calculate the mass and position of each and every planet and moon with tremendous precision. However, they will encounter the problem that they can never cease this planetary game of billiards because, if control is relinquished, calamities are unavoidable.

As chaos exists, an advanced civilization wanting to change planetary orbits may perhaps find a smarter method than the one I described at the beginning of this chapter. They will understand that it makes no sense to sling asteroids between Venus, Mars and Jupiter like huge swings, but instead they will send their asteroids to Pluto to disrupt its orbit. Pluto will then over time cause pre-calculated changes to Mars's and Venus's orbits.

They will probably also want to reposition the Earth. Astronomers who have studied the Sun know that it will grow bigger and brighter in the coming hundreds of millions of years. And hotter too. By that time, we want the Earth to be a bit further away from the Sun. Obviously, a civilization that has such ambitious plans would have to be able to do what we cannot; recognize the sense and sensibility of making investments that will only reap rewards in millions of years. These civilizations will require a political system that is totally different from ours!

chapter 18

Idiocracy

IF YOU THOUGHT that the previous chapter was not eccentric enough, then I advise you to take any science fiction book off the shelf, as most of these authors write about much more unrealistic or even completely impossible notions. Along these lines, Ed Regis describes fabulous futuristic fantasies in his book *Great Mambo Chicken and the Transhuman Condition: Science Slightly over the Edge*. People who see their death approach, either because they are terminally ill or are of old age, have their bodies frozen and stored, in the expectation that science will be able to de-ice and heal them. But will the delicate refrigeration procedure be carried out correctly? Can we trust future civilizations to willingly defrost anyone? Won't they say, "*too bad, but you are permanently damaged and we will send your body into space with the rest of our garbage*"? Regis calls this unbridled optimism 'hybris', a mixture of extreme arrogance and megalomania with gullibility.

Others expect a future in which the human brain will become entangled with electronic machines in such a way that the human race will continue to exist as hybrid robots. All aging processes will be halted because of new scientific discoveries and no one will die. Then these half humans — half robots will conquer the universe and, whereas I have described the practical scientific limitations of travelling at roughly 1,000 kilometers per second, these beings won't be restrained by such a puny little obstacle as the speed of light. Eventually, the 'point Omega' is reached, whereby human consciousness merges with the entire cosmos.

Inspired by science fiction series such as Star Trek, we make giant leaps into superspace and our spaceship is slung from one planet to another with warp speed, ignoring all laws of relativity theory and unhindered by any logical reasoning whatsoever. In my world, a voyage over such distances takes tens of thousands of years and even then you haven't got that far in interstellar or galactic terms. It is simply not possible to transport people with velocities anywhere close to the speed of light, because of the astronomical energy costs, because of the deadly radiation unavoidable in such a situation, and because the expenses of such a trip are not worth it.

Humans therefore won't be able to do it, but perhaps biological life can. Our neumannbots might want to take refrigerated bacteria and algae along with them, as well as all the required information to start higher life forms once they have arrived on a suitable planet. In this way, we might be able to spread biological life, if we were so inclined.

A second idiocracy I encounter over and over again concerns communication. In many science fiction stories, to enable contact with civilizations in other galaxies, it is assumed that communication faster than the speed of light will be possible. According to strict natural laws, however, every message would take many years before it is answered; hundreds of years even for star systems that are not that far away. No fast-paced adventure story is able to accommodate this complication in its story line! Too bad.

Another popular way out is even cheaper: why don't we use our telepathic or other psychic powers? Our world knows many paragnostics and other paranormally 'gifted' individuals, and we hear so much about them that we are almost tempted to believe in their special powers: apparitions and phenomena that are entirely unaffected by the limitations set by natural laws. We could simply communicate through paranormal channels, and because paragnostics have little understanding of relativity theory, even this is not considered a serious obstacle.

I have time and time again insisted that all these purported phenomena exist solely in our imagination, that no one has paranormal abilities, and that information, however inconsequential, can only be exchanged according to the laws of Nature. Regularly I am laughed and jeered at, but I maintain I am right, not in the least because of the bet I have put on my own website, inviting any paranormal who believes he or she has special powers to take it on. They only have to pass one test, for which the conditions have been formulated by me, and I believe that the numerous assertions made regarding paranormal occurrences will easily rise to the challenge if there is even an inkling of truth to them. No one has yet responded to my invitation. I am sorry if you are one of those believers. I am not under the illusion that I can convert anyone, but I will not be sympathizing either.

The world famous illusionist James Randi even has an amount of one million dollars ready to be given to anyone who can convince him of his or her paranormal powers. No one has yet been able to collect. Most claims were simply too vague to be tested accurately at all.

Would mega projects with new scientific and technological dimensions be able to put humanity on a higher plane? Will there ever be any human colonies thriving on neighboring planets and moons? Will any of our machines ever reach nearby stars? And will we be able to slightly change the Earth's orbit if the Sun bursts at the seams?

A while ago I was attending a conference in Copenhagen and had a free afternoon to wander around town. Deep in thought, I was strolling along the water when I walked past the world famous sculpture of a mermaid. A mutation? A genetic manipulation? Are we dealing with people here who wanted to live an aquatic existence? Further along in the harbor a large ship was docked. Only when I neared it, did I realize how gigantic this vessel really was — it was a cruise ship that must have completed numerous voyages, an old-fashioned passenger carrier. Only then I noticed the proud lettering on the ship's bow: Statendam!

That was one of my father's ships, an ocean steamer, a floating city at sea! I still have a picture of my mother on that same ship, from 1957. And all of a sudden I saw how the future must have looked fifty years ago: we dreamed of increasingly large vessels that would give our life new dimensions. Every era generates its dreams and there have always been those who can make them come true, as long as they comply with the laws of engineering; the immutable and beautiful laws of physics.

Websites*

More information about subjects in this book can be found at the following websites:

Moore's Law:
http://www.intel.com/technology/mooreslaw/

The Maasdam:
http://www.destinationoceans.com/cruise-resources/lines/holland-america-line/maasdam.cfm

Professor Frits Schoute's ecoboat:
http://www.ecoboot.nl

Zero Emission Research and Initiatives, reforestation of the savanna in Eastern Colombia:
http://www.zeri.org

---

*Please note that websites tend to be volatile; by the time this book has appeared several of these websites may no longer exist. The reader is then advised to search for similar sites. They are usually not difficult to find, with modern ICT technology.

The Earth as a snowball:
http://en.wikipedia.org/wiki/snowball_earth

The Laddermill:
http://www.ockels.nl

The Pioneer 10:
http://www.nasa.gov/mission_pages/pioneer/index.html

The Moon Hotel:
http://www.rombaut.nl/

The Mariner Flyby:
http://home.earthlink.net/~nbrass1/mariner/miv.htm

Space Colonies:
http://space.mike-combs.com

The Cassini-Huygens:
http://saturn.jpl.nasa.gov/home.index.cfm

The Kuiper Belt and the Oort Cloud:
http://en.wikipedia.org/wiki/oort_cloud

Homepage of the author:
http://www.phys.uu.nl/~thooft

The James Randi Educational Foundation
(the Million Dollar Challenge):
http://www.randi.org

Illustrations

The publisher has attempted to identify all rightful owners. However, anyone else who claims any rights is requested to contact World Scientific Publishing Co., editor@wspc.com.sg.

p. 63:   Courtesy of Gunter Pauli, Zero Emissions Research and Initiatives (ZERI)
p. 106: Courtesy of Space Telescope Science Institute, NASA

For further information:

http://www.zeri.org
http://hubblesite.org/newscenter/archive/